Social Sciences: a second level course
New trends in geography   Units 3–4
prepared by the course team

4

# Economic geography — industrial location theory

The Open University Press

The Open University Press
Walton Hall Milton Keynes

First published 1972. Reprinted 1973

Designed by the Media Development Group of the Open University.

*Printed in Great Britain by*
EYRE AND SPOTTISWOODE LIMITED
AT GROSVENOR PRESS PORTSMOUTH

ǀ SBN 0335 01531 X

This text forms part of the correspondence element of an Open University Second Level Course. The complete list of units in the course is given at the end of this text.

For general availability of supporting material referred to in this text, please write to the Director of Marketing, The Open University, P.O. Box 81, Milton Keynes, MK7 6AT.

Further information on Open University courses may be obtained from the Admissions Office, The Open University P.O. Box 48, Milton Keynes, MK7 6AB.

# Contents

## Introduction to Block II

# Introduction to Block II

This block, which represents six weeks work, is designed to introduce you to the basic concepts of economic geography. Space permits a consideration of only those factors which are relevant in explaining the location of industry, although the concepts discussed can be applied throughout economic geography.

On a world scale great concentrations of industry can be observed both in the Manufacturing Belt of the USA and in western Europe. Unit 3 describes these concentrations. It also reviews the sort of theories which have been put forward since 1900 to explain location patterns. Units 4–7 consider the factors which help to determine the location of factories. Unit 4 examines how the cost of labour, transport or raw materials, which a firm needs in order to produce its goods, vary from place to place. An industrialist anxious to reduce his production costs will naturally locate in those areas where the factors of production are cheapest. Unit 5 examines the impact of government intervention on location decisions and considers the way in which the demand for a firm's products may vary from place to place. Unit 6 considers how the volume of production of a factory and the way in which the factory is organized can influence costs. It also discusses the way in which managers adjust to changes in the economic and social environment. Unit 7 focuses on the way in which industrial managers make location decisions and the extent to which they take into account the factors discussed in the previous parts. Unit 8 is a study of the British iron and steel industry and is an attempt to show how the theoretical examination of location patterns can be illuminated by empirical studies.

---

Books which will be of use to you in working through the block are as follows:

## Set Book

Chorley, R. J. and Haggett, P. (1968) *Socio-economic Models in Geography*, London, Methuen.

## Recommended Books

Chisholm, M. (1970) *Geography and Economics* (2nd ed.), London, G. Bell & Sons.
Estall, R. C. and Buchanan, R. O. (1966) *Industrial Activity and Economic Geography* (rev. ed.), London, Hutchinson University Library.
Estall, R. C. (1972) *A modern geography of the United States*, Harmondsworth, Penguin Books.
Hoover, E. M. (1948) *The Location of Economic Activity*, New York, McGraw-Hill.

Karaska, G. T. and Bramhall, D. F. (eds.) (1969) *Locational Analysis for Manufacturing: A selection of readings*, Cambridge, Massachusetts, MIT Press.

McCarty, H. H. and Lindberg, J. B. (1966) *A Preface to Economic Geography*, Englewood Cliffs, Prentice-Hall.

Smith, D. M. (1971) *Industrial Location: An economic geographical analysis*, New York and Toronto, John Wiley & Sons.

Smith, R. H. T., Taaffe, E. J. and King, L. J. (eds.) (1968) *Readings in Economic Geography. The Location of Economic Activity*, Chicago, Rand McNally.

Smith, W. (1968) *An historical introduction to the economic geography of Great Britain* (1st edition reprinted with an appreciation by M. J. Wise), London, G. Bell & Sons.

Chapter 10 'Models of Industrial Location' by F. E. Hamilton in Chorley and Haggett (1968) *Socio-economic Models in Geography* provides a useful review of the location theories dealt with in Unit 3 of the block. Chisholm (1970) *Geography and Economics* is a fairly straightforward introduction to some of the main concepts in economic geography as is Estall and Buchanan (1966) *Industrial Activity and Economic Geography*. Both of these should prove useful reading throughout the block. Estall (1972) *A modern geography of the United States* provides more information on the location of manufacturing industry in the USA and is useful for Unit 3. Hoover (1948) *The Location of Economic Activity* is an example of the location theories outlined in Unit 3. It also discusses freight rates which are discussed in Units 4 and 8, and government policy to industry which is discussed in Unit 5. Karaska and Bramhall (1969) *Locational Analysis for Manufacturing* is a collection of papers on various aspects of industrial location theory and contains sections which are relevant to all six units. McCarty and Lindberg (1966) *A Preface to Economic Geography* is a rather more advanced and comprehensive introduction to economic geography than either Chisholm or Estall and Buchanan. *Industrial Location* by D. M. Smith is a fairly advanced text. It is particularly strong on quoting empirical examples of how costs of production and demand vary from place to place. It also provides a very clear review of the location theories discussed in Unit 3. *Readings in Economic Geography* by R. H. T. Smith *et al.* is a collection of papers which will be found useful for Units 3 and 8 of the block. W. Smith (1968) *An historical introduction to the economic geography of Great Britain* is again useful for Units 3 and 8. The original issue of Smith's book is to be preferred but is out of print. It appeared in 1949 as *An Economic Geography of Great Britain*. The reprinted text is only the first part of this book.

The parts should give you a fairly good knowledge of the major concepts in economic geography as defined in the block introduction. However, if you have time to do further reading you should concentrate first of all on Estall and Buchanan and then to on go study D. M. Smith (1971) *Industrial Location* and those papers in Karaska and Bramhall to which reference has been made. The other books listed can be consulted as time permits.

# Unit 3
# Concentration and dispersal

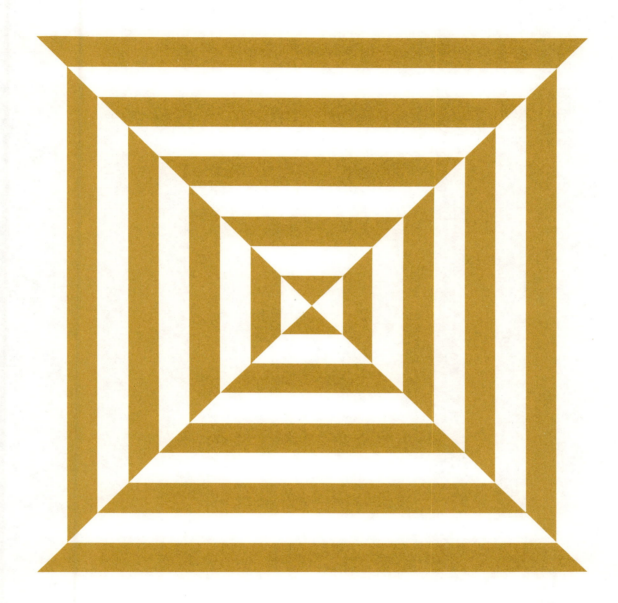

prepared by Andrew Blowers
for the course team

# Unit 3 Contents

## A   Patterns of economic activity

# 1   Introduction

Economic geography is about the distribution of man's economic activities on the surface of the earth. At whatever scale we examine these activities, we are immediately aware of the complexity of their distribution. It is the geographer's task to unravel this complexity and to *discover* and *describe* the patterns of distribution, and to *explain* the processes responsible for them. That patterns do exist and that there is an underlying order in them will be demonstrated in this unit. We shall show that it is possible to comprehend these patterns by developing theories and by testing them against the real world patterns presented by empirical studies.

## 1.1   The sectors of the economy

Economic activities may be broadly categorized into three groups, primary, secondary and tertiary. *Primary activities* are those by which man gains products from the resources of the natural environment. He thus produces food and certain raw materials by means of agriculture and extracts minerals by means of mining. *Secondary activities* are those devoted to manufacturing and it is with these that this block is mainly concerned. The processes involved may vary from the painstaking manual labour of domestic craftsmanship to the highly specialized technology of mass production. In the modern industrial state increasing specialization has resulted in a complex chain of interlocking and interdependent manufacturing activities, relatively few of which are producing articles for final consumption. The size of manufacturing enterprises varies. On the one hand there are small firms engaged in a single process located in one plant. By contrast the scale of investment and the degree of organization required in many activities is increasing, and has often led to the concentration of control and the emergence of large international corporations employing thousands of workers on a wide range of products in hundreds of plants. Differences in the size and organization of industrial enterprises has important locational implications as we shall show later in this block. There is a third group, the *tertiary or service activities*, which are essential for the organization, control, and development of societies. These include the administration of central and local government and the panoply of social services, which in the UK are provided by the state, such as health, welfare, and education. Another group of services are those required for the efficient economic functioning of society such as commerce, retailing, finance, insurance, and transportation. Construction, professional and cultural services, and defence are further examples of activities commonly regarded as falling within this group. Using these three broad categories we may discern three distinct stages in economic development.[1] There is, first, an agricultural phase when the economy is dominantly rural, labour intensive, and technically backward. The second or industrial phase is marked by an intensification of economic activity, increasing specialization, and the diversification of activities. Manufacturing becomes the major economic sector but agriculture benefits from improved

---

1   At this point you might like to refer to the chapter on 'Models of Economic Development' by D. E. Keeble in Chorley and Haggett (eds.) (1968) *Socio-economic models in Geography*, Ch. 8, especially pages 243–54.

techniques of production which lead to higher productivity and a consequent decline in its labour force. The third and latest stage of evolution is the expansion of the tertiary sector until it becomes the most important in terms of output and income.

The so-called advanced nations of the world have reached this third stage but there are many economies which have yet to achieve the 'take off' into the self-sustaining economic growth postulated by Rostow (1967). In some countries a high proportion of the population works on the land. In India, for example, at least two-thirds may be dependent on the agricultural sector (Spate and Learmonth, 1967, p. 121). By contrast in the prosperous and developed parts of the world the agricultural population is quite small. In the UK for example it is as low as 1·7 per cent since we import about half our food and most of the agricultural raw materials that we need. The USA which is largely self-sufficient has 4·9 per cent in the agricultural sector. The primary producers are frequently and erroneously regarded as synonymous with the developing countries. Yet Australia, New Zealand, South Africa, and Canada which have only a small proportion of their employed population engaged in the primary sector derive much of their export earnings from primary products. A substantial proportion of the world's trade in food, minerals, and other raw materials in fact emanates from the advanced countries. In these countries, however, the numbers engaged in manufacturing industries is relatively high. It is this that distinguishes their economic structure from that of the developing countries. Although the developed countries commonly have more than half the employed population in the tertiary sector this may also be the case in some developing countries. Table 1 which shows the population employed in each of the three economic sectors for selected countries illustrates these differences in economic structure between countries. It may be looked at in conjunction with the illustration of Rostow's model in Chorley and Haggett (1968) *Socio-economic models in Geography*, page 251.

**Table 1**
Economic structure of selected countries. Percentage of working population engaged in the three economic sectors.

|                    | Primary | Secondary | Tertiary |
|--------------------|---------|-----------|----------|
| UK (1969)          | 3·4     | 38·8      | 57·7     |
| France (1968)      | 17·0    | 28·0      | 55·0     |
| USA (1969)         | 5·7     | 27·3      | 67·0     |
| UAR (1967)         | 50·6    | 11·0      | 38·4     |
| Gabon (1968)       | 25·7    | 10·9      | 63·4     |
| Puerto Rico (1969) | 11·9    | 19·7      | 68·4     |

Source: International Labour Organisation.

## 1.2 Regional specialization and trade

One of the spatial consequences of economic development is the increasing regional specialization of production and the growth of trade.[1] We may study this specialization and trade at various spatial scales. At a microspatial level we may focus upon the industries and the linkages between them within, say, a city or a region like north-east England. At a national level we are concerned with regional economies and the development of interregional trade flows.

1 There are also demographic consequences which have a spatial context. One is the increase in urbanization and another is the increase in migration both between and within cities.

Finally, at an international level it is national economic development and the pattern of international trade that demands our attention.

At the international level this pattern of specialization and trade may be conceptualized in terms of *comparative advantage*. The concept is a familiar one in economics and is developed in most basic text books in the subject.[1] It is based upon the fact that different economies have different endowments of the factors of production. The costs of using these factors to produce particular commodities may be measured in terms of the alternative uses to which they might otherwise be put. So, for example, the cost of producing a car may be expressed in terms of, say, 100 prams, 25 bicycles, or half a cabin cruiser. Such a measurement is known as the *opportunity cost*, that is, the benefit foregone by using factors in a particular way. Since countries have different factor endowments it follows that the opportunity costs of producing a set of commodities will vary. Some countries will have lower opportunity costs in certain commodities than in others. In these commodities they are said to have a comparative advantage. By specializing in these commodities and exchanging them for other commodities which it needs (and in the production of which other countries have a comparative advantage) a country can increase its total production. Thus, so the theory goes, international trade is mutually beneficial to all participants, whatever their productivity. The important point to notice is that it is comparative and not absolute advantage that matters.

Although free trade may, ultimately, lead to higher incomes for all participants there are often sound reasons for maintaining some protection. This is particularly so when the assumptions of the theory are not fulfilled. For example, where relative prices do not reflect comparative costs free trade may lead to a misallocation of resources. Again, full employment which is assumed in the theory may be imperilled by a re-allocation of resources on the basis of free trade. In any case a major shift in production may be resisted by the labour force. Indeed, the need to reduce unemployment may be held as a justification for some measure of protection. Countries may also argue that protection enables them to nurture infant industries from which considerable scale economies may accrue once they are fully developed. Protection may also be invoked on strategic grounds or as a means of reducing dependence on a narrow economic base.

When we come to the smaller scale of the nation, we may, again, explain regional specialization on the basis of relative factor endowments leading to differential opportunity costs. Thus, as Chisholm demonstrates, the regional concentration of the cotton industry is to be explained by the fact that at the time it was established there were fewer opportunities foregone in Lancashire than in, say, Clydeside or South Wales (Chisholm, 1970, pp. 23–6). There is, however, an important difference between international and interregional specialization. At the international level the movement of factors such as capital and labour across frontiers is strictly limited.[2] As a result factors may have different productivities and different earnings but benefits may still accrue from trade. Within a country or economic bloc there are no such barriers and factors are relatively mobile. If there were perfect mobility of factors labour would migrate to areas of high wages and capital (i.e. investment) would move to areas of lower costs. Eventually a state of equilibrium would be achieved

1   An explanation of the concept was given in D100 Unit 3 *The Economic Basis of Society* of the Foundation Course *Understanding Society*, pp. 66–70. It is given a geographical context by Chisholm (1970) pp. 18–28.

2   We are obviously concerned with economic and not political frontiers. Thus a trading bloc such as the EEC allows considerable mobility of factors, especially labour, across the national frontiers of its members.

where, through a process of constant adjustment the earnings of all the factors of production are equal. It is precisely because tariff barriers are absent that those regions which are poorly endowed will be unable to specialize in those commodities where they have a comparative advantage and will tend as a result to decline. Thus, within a national economy interregional specialization tends to develop on the basis of *absolute advantage*. In reality, of course, factors are far from perfectly mobile and so regional inequalities tend to persist.

## 1.3 Differences between agricultural and industrial distributions

When we come to look at the spatial pattern of economic activities on the world scale we are immediately struck by the contrast between industry and agriculture. Industry is highly concentrated geographically, whereas agriculture is a relatively land-extensive activity. Agricultural types and the unit of production, the farm, occupy tracts of land. The spatial pattern they form we may, therefore, describe as an *areal* one. Our interest in agriculture lies in analysing the distribution of types of agriculture and in the changing pattern of agricultural land use. Increasingly, agricultural geographers have turned their attention to the agricultural system as a whole and its relationship to other economic activities. Models have been developed as an aid to explanation. Some geographers have begun to examine the decision making processes which have a spatial context (e.g. the diffusion of agricultural innovations[1]). Lack of space prevents us from tackling agricultural geography in this block but those of you who are interested could read as an introduction to the subject, Chorley and Haggett (1968) Chapter 11.

Modern industry, for efficient operation requires the integration of machinery and a labour force within a plant or factory. By increasing the size of the plant it may be possible to achieve internal economies of scale. Furthermore, a close association of plants may bring about changes in the external economic environment such as a large labour pool, a variety of entrepreneurial services, and a developed infrastructure (transport, housing etc.). As a result of this geographical concentration, industries will be able to achieve external economies of scale (sometimes called agglomeration economies). The internal cost structure of firms and agglomeration economies are examined in Unit 6 of this block. For the moment we need merely note the tendency of industries to cluster. The pattern which they adopt may be *punctiform* (where clustering occurs around a series of points) or they may be diffused over a wide area in which case they may form an *industrial belt*. In either case our interest lies in explaining these patterns whether we focus our attention on the world scale or on that of the individual factory.

## 1.4 Industrial belts of the world

If we look at the map of industrial belts on the world scale (figure 1), we may observe that industry has tended to concentrate into a number of belts. Outside these belts industries, though widely scattered, have concentrated at particular points. There are several major belts, the biggest of all being that of the north-eastern United States. This area, contained within the quadrilateral linking Boston-Minneapolis-St. Louis-Washington (well over a thousand miles at

1 This subject is dealt with, to a limited extent, in the Foundation Course *Understanding Society* (Radio Programme 19) and, in more detail, in the Agricultural Block (Block III) of the Second Level Course D203 *Decision Making in Britain*.

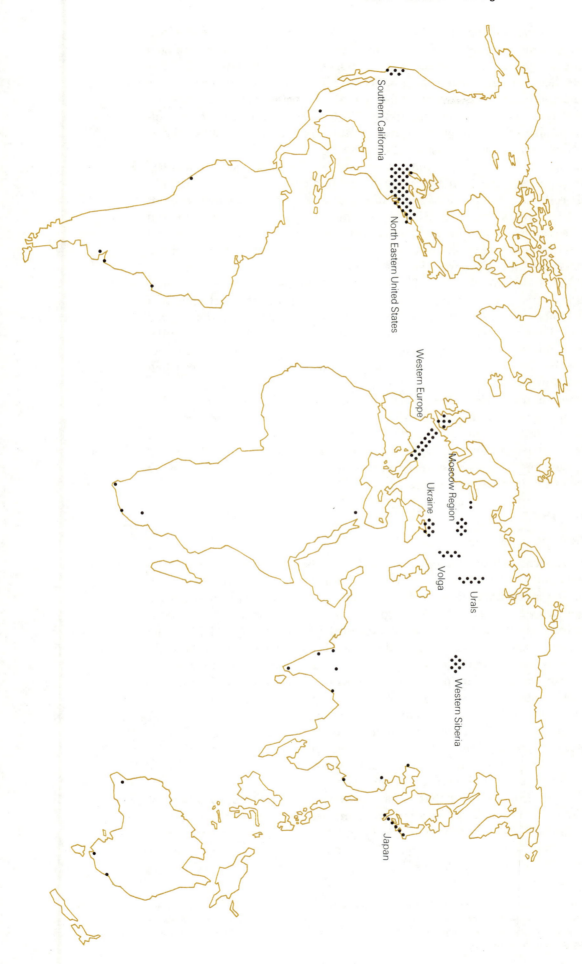

**Figure 1**
Industrial belts of the world

its widest) contains about two thirds of the industrial activity of the USA. In the remaining parts of the country only southern California has managed to sustain a wide range of industrial activities. A possible explanation of this lies in the fact that only California is sufficiently insulated by distance from the north-east and has a large enough population to develop a diversified industrial structure. This argument may be applicable to certain industries but is insufficient as an explanation of the overall industrial pattern. Such an explanation is, of course, the main purpose of this block.

In Europe, an axial belt, albeit discontinuous, stretches from Manchester to Milan (a linear distance of about 800 miles). Somewhat similar in geographical scale is the industrial belt of the southern coastline of Japan which extends for about 600 miles. In the USSR it is possible to define six major industrial areas. The region surrounding Moscow is the largest, and Leningrad, the second city, forms the centre of another industrial region. The industry of the Donbass in the Ukraine and of western Siberia is more scattered as these areas were initially based on coalfields. The Urals and Volga industrial regions comprise a series of scattered industrial towns. It is difficult to grasp the vast scale of a country which is equivalent in area to ninety times the size of the United Kingdom. Thus an area like the Urals which stretches for a thousand miles from north to south and which, on a Russian scale, is a relatively compact region would be large enough to encompass the entire manufacturing belt of western Europe.

Elsewhere in the world industries tend to be focused around a single conurbation. Examples are the major cities of Australia, India, and Latin America.

So far we have concentrated on the global pattern. We shall now look in more detail at two individual regions and attempt to explain their industrial location pattern. The regions we have chosen are the USA and the UK as these are examples of different areal scales.

## 2 The USA[1]

### 2.1 Regional industrial patterns

As we have seen, industrial activity in the USA is heavily concentrated in the Manufacturing Belt of the north-east with southern California constituting a second major industrial region (see Figure 2). By using some simple quantitative measurements we can give a greater degree of precision to our statements about the locational pattern. The measurements used here are all based upon the location quotient, which was introduced in Block I, and its derivatives. Basically it is a device whereby we can describe the relative concentration of a particular phenomenon, in this case industrial activity. It should be emphasized, however, that statistics must be interpreted with great care and that 'they are essentially mechanical devices with which empirical facts can be processed to reveal certain statistical tendencies or regularities'. (Isard, 1960, p. 166.) You must hesitate to draw conclusions that are not really supported by the data available. You should always look critically at the data, their relevance to the particular problem, and the methods used to process them.

It is evident from Table 2 that certain regions may have concentrations of particular industries (i.e. the location quotients are high). For example, New England (for the location of regions see Figure 3), has a higher proportion of workers (or, measured another way, of value added by manufacture), in the textile, paper, rubber and plastics, leather goods, and machinery and instru-

1 In the recommended text by Smith, Taafe, and King (eds.) (1968) there are two papers on the geographical structure of industry in the USA. These are Pred, A. 'The Concentration of High Value Added Manufacturing', pp. 158-72 and Harris, C. D. 'The Market as a Factor in the Localization of Industry in the United States', pp. 186-99.

Figure 2
Manufacturing regions of the USA

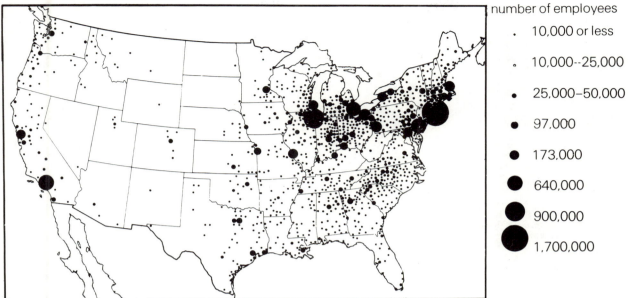

number of employees

.    10,000 or less

°    10,000--25,000

•    25,000–50,000

●    97,000

●    173,000

●    640,000

●    900,000

●    1,700,000

Source: J. W. Alexander, *Ecomomic Geography*, Prentice Hall, 1963

ments industries than would be the case if each industry were directly related to the distribution of the total working population. Conversely the Mountain States have location quotients higher than 1 in the food, lumber, printing and publishing, stone, primary metal, transportation and ordnance industries. Such variations provide a rationale for interregional trade, if certain conditions, such as those we discussed earlier, are fulfilled.

At the scale of the individual city the contrasts are much greater. The reason for this is obvious – the smaller the unit considered the less likely is it to correspond to the national average. Ultimately we should be considering individual cases, all of which are different in some way or another. This can be demonstrated by the case of Flint, Michigan, which has a relatively high location quotient (2.8) for the transportation industry. This is a case of specialization in one industry. The location quotient for the transportation industry for the city of Detroit (2.7) is slightly lower and that for the East North Central region is lower still (1.4) reflecting its more diversified industrial structure.[1]

## 2.2   The distribution of individual industries

The overall pattern for individual industrial sectors may be examined by means of the *coefficient of localization* (or *geographical association*) and the similar *index of concentration*. Industries with high coefficients have a concentrated pattern relative to the distribution of the population as a whole. Among these are (a) the tobacco, timber, and petroleum industries which are related to localized raw materials; (b) interdependent industries such as instrument makers and specialized machine tool manufacturers; (c) industries based upon cheap labour pools of which textiles is the most important example; (d) specialized industries dependent upon large consumer markets, such as New York (e.g. publishing); (e) industries where economies of scale may be achieved in large plants (e.g. aircraft). By contrast industries dependent upon ubiquitous raw materials, or those whose transport costs are high in relation to total production costs, such

1   These patterns of local and regional specialization may also be expressed graphically by means of a *specialization (or diversification) curve* based upon the principle of the Lorenz curve as explained in the Course Notes. The Lorenz curve was also introduced to demonstrate personal income distribution in the Foundation Course *Understanding Society* (Radio Programme 16).

Table 2

US Location quotients of industries by regions (1963)

| major industry group | regions | | | | | | | | | total in thousands |
|---|---|---|---|---|---|---|---|---|---|---|
| | NE | MA | ENC | WNC | SA | ESC | WSC | M | P | |
| food, kindred products | 0·50 | 0·75 | 0·77 | 2·01 | 0·98 | 1·03 | 1·57 | 1·85 | 1·21 | 1,715 |
| tobacco manufacturers | 0·05 | 0·55 | 0·11 | 'D' | 4·84 | 2·69 | 0·10 | — | 0·01 | 84 |
| textile mill products | 1·36 | 0·69 | 0·11 | 0·43 | 4·41 | 1·64 | 0·21 | 0·04 | 0·10 | 883 |
| apparel, related goods | 0·77 | 1·83 | 0·28 | 0·65 | 1·28 | 1·99 | 0·94 | 0·32 | 0·53 | 1,291 |
| lumber, wood products | 0·58 | 0·24 | 0·39 | 0·58 | 1·54 | 2·20 | 1·90 | 2·99 | 2·75 | 569 |
| furnitures, fixtures | 0·67 | 0·71 | 0·88 | 0·58 | 1·95 | 1·69 | 1·09 | 0·51 | 0·93 | 381 |
| paper, allied products | 1·33 | 0·90 | 0·92 | 0·87 | 1·05 | 0·95 | 1·12 | 0·33 | 0·85 | 614 |
| printing, publishing | 0·90 | 1·21 | 0·93 | 1·41 | 0·72 | 0·63 | 0·95 | 1·33 | 0·90 | 926 |
| chemicals, allied products | 0·43 | 0·94 | 0·68 | 0·76 | 1·30 | 1·46 | 1·47 | 0·56 | 0·54 | 853 |
| petroleum, coal products | 0·11 | 0·49 | 0·50 | 0·63 | 0·18 | 0·24 | 4·83 | 1·50 | 0·90 | 220 |
| rubber, plastic products | 1·92 | 0·78 | 1·39 | 0·52 | 0·83 | 0·51 | 0·95 | 0·67 | 427 | 427 |
| leather, leather products | 3·47 | 1·18 | 0·54 | 1·51 | 0·40 | 1·18 | 0·57 | 0·53 | 0·18 | 335 |
| stone clay, glass products | 0·50 | 0·90 | 0·94 | 1·11 | 1·10 | 1·01 | 1·36 | 1·56 | 0·86 | 603 |
| primary metal industries | 0·61 | 1·12 | 1·48 | 0·42 | 0·52 | 1·04 | 0·65 | 1·30 | 0·51 | 1,167 |
| fabricated metal products | 1·00 | 0·93 | 1·37 | 0·86 | 0·51 | 0·83 | 0·92 | 0·59 | 0·85 | 1,111 |
| machinery, except electrical | 1·23 | 0·82 | 1·54 | 1·24 | 0·31 | 0·44 | 0·76 | 0·61 | 0·76 | 1,521 |
| electrical machinery | 1·27 | 1·09 | 1·08 | 0·74 | 0·49 | 0·57 | 0·53 | 0·56 | 1·15 | 1,612 |
| transportation equipment | 0·84 | 0·51 | 1·40 | 1·06 | 0·56 | 0 46 | 0·84 | 1·29 | 1·57 | 1,690 |
| instruments, related products | 1·95 | 1·54 | 0·86 | 0·85 | 0·33 | 0·21 | 0·44 | 0·53 | 0·70 | 316 |
| miscellaneous manufacturing | 2·01 | 0·50 | 0·76 | 0·88 | 0·43 | 0·55 | 0·52 | 0·56 | 0·67 | 398 |
| ordnance, accessories | 0·73 | 0·14 | 0·30 | 1·33 | 0·45 | 0·54 | 0·99 | 3·27 | 6·66 | 248 |
| total in thousands | 1,425 | 4,075 | 4,483 | 1,014 | 2,125 | 892 | 865 | 284 | 1,799 | 16,961 |

Location quotient = $\dfrac{\text{Region's employment in industry A as percent of national employment in industry A}}{\text{Region's employment in all industries as percent of national employment in all industries}}$

'D' – withheld to avoid disclosing figures for particular companies
Figures in brown denote location quotients greater than 1·00

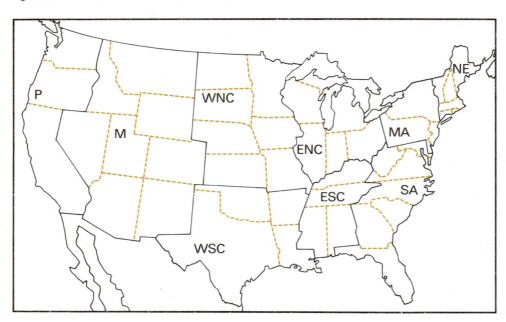

| NE | New England |
|---|---|
| MA | Middle Atlantic |
| ENC | East North Central |
| WNC | West North Central |
| SA | South Atlantic |
| ESC | East South Central |
| WSC | West South Central |
| M | Mountain |
| P | Pacific |

----- state boundaries
——— census regions

Figure 3
USA census regions

as food, furniture, paper and printing, and chemicals[1] have a dispersed pattern corresponding fairly closely to the distribution of population. They have low coefficients and the variations between them may be illustrated by the *localization* curve.

## 2.3 Changes in the distribution of industry

The measurements described so far are static ones. We may add a further dimension by introducing a dynamic element, that is by analysing trends over time. There are several ways in which this may be done. Basically the principle of such measurement is to show the rate of change of industrial growth within a region relative to changes in the country as a whole. Victor Fuchs with the help of census data has devoted a whole book to the detailed consideration of such comparative changes in the regional location of industry in the USA between 1929 and 1954 (Fuchs, 1962a, 1962b). From Table 2 it is clear that there has been a tendency in the USA for regions to become somewhat more alike in their industrial structure. Despite this evidence of some convergence great disparities still exist between regions.

The most striking change in the distribution of industry in the United States has been the increase of industry in the south and west relative to the rest of the country. Fuchs shows that the south and west had a quarter of the employees in manufacturing and one-fifth of the value added by manufacture in 1929 but that by 1954 the proportions had risen to one-third for both indices. Such a regional shift of industry may, in part, be explained by *differences in the industrial structure of various regions*. Thus areas such as New England, parts of the south, and the Mountain States had an industrial structure in which slowly growing industries were relatively important. Other parts of the Manufacturing Region especially the western part, and California in particular, had structures with a high proportion of rapidly growing industries. These structural factors undoubtedly played a part but Fuchs concludes that *differential regional growth rates within the various industrial sectors* offer a more convincing general explanation for the overall change in the regional distribution of industry that has occurred. Thus industry has grown in the south despite its poor overall structure. Differential growth rates between regions reflect different rates of expansion or contraction of particular industries, different birth and death rates of plants, and (probably more rarely) the relocation of factories from one region to another.

The reasons for this differential growth of industry appear to be mainly those which affect the supply of the factors of production. Fuchs, from the evidence for the period 1929–54, reckoned that one-third of the change (or shift) in manufacturing could be accounted for by the availability of natural resources in the south and west (and he included the attractiveness of their climate under this head) and another third by the supply of cheap labour. Thus growth has been especially rapid in California and Florida where climate and amenity factors have been important, in Texas owing to its resources of oil, and in parts of the south where cheap labour had attracted decentralizing industries such as textiles, (although the cotton industry is now much more capital intensive). In sparsely populated regions such as the Mountain States only those industries serving the small local market or based upon local raw materials are represented. This is because most other industries require a certain population size (or 'threshold') from which they may draw labour and services before they can become established.

Fuchs's emphasis on the supply of the factors of production seems to refute the common assertion that industrial shifts can be explained largely in terms of

---

1  In the case of chemicals, this statement applies to the industry as a whole. Certain types of chemicals will be concentrated, e.g. organic chemicals, based on the oilfields of Texas and Louisiana.

population and income (i.e. demand). He argues that 'it would be at least as plausible to suggest that the shifts in manufacturing were responsible for changes in population and income' (Fuchs, 1962a, p. 88). Whatever the explanation for the industrial redistribution within the USA, the concentration of industry within the Manufacturing Belt still remains the dominant feature of its industrial geography.

### 2.4 Current trends and the future pattern

A conflict between centripetal and centrifugal forces is evident at the local scale. Although the major cities continue to be industrial centres there is a tendency for industries to decentralize to the suburbs in search of cheaper and more abundant land. Thus although the population as a whole has become more concentrated as a result of urbanization, there has been a tendency for the urban population itself to become increasingly diffused.

Zelinsky, who like Fuchs made a detailed statistical analysis of post-war locational changes confirmed the long-term trend towards industrial dispersal, although on the local scale it appears to have been interrupted during the period 1939–47. (Zelinsky, 1962). He found a positive correlation between the rate of deconcentration and the size of cities. However, although industry had decentralized from the major cities the most rapid growth was within the metropolitan ring surrounding them and thus manufacturing has remained an urban phenomenon.

Tertiary activities remain concentrated at the urban core, though they too, have begun to decentralize. Brian Berry, speculating on the locational pattern of the United States in the year 2000 predicts an inversion of the current situation. This, he suggests, will come about as a result of electronic forms of communication reducing the need to overcome distance. 'Traditionally we have moved the body to the experience; increasingly we will move the experience to the body and the body can therefore be located where it finds the non-electronic experiences most satisfying' (Berry, 1970, p. 49). This will be, according to Berry, places of high amenity areas which are at present peripheral to the major centres of activity. Of course, their popularity will eventually threaten their amenity. It is unlikely that the pattern of the year 2000 will correspond closely to Berry's predictions and certainly locational freedom is unlikely to be available to everyone. However, by attempting to interpret the future he reminds us that the future pattern is unlikely to be a mere extrapolation of existing trends but will be shaped by locational decisions now and in the future.

Whatever the future pattern, the present is the outcome of past decisions and we may now turn to the UK to explore the locational pattern in its historical context. Both in areal scale and in the direction of regional shift the UK offers some interesting contrasts to the USA.

## 3 The United Kingdom[1]

### 3.1 The industrial pattern

If we look at a population map of the UK (Figure 4) we are immediately aware of the unevenness of the distribution. There is a belt of high density running diagonally across the country from the north-west to the south-east and including five of the seven major conurbations (Greater London, the West Midlands, Greater Manchester, the West Riding and Merseyside) and a number

---

1 Useful background reading for this section on the UK will be found in the recommended text by Smith, W. (1968) *An Historical Introduction to the Economic Geography of Great Britain*, London, G. Bell & Sons.

of large cities (Sheffield, Nottingham, Coventry, Leicester and so on). This belt was dubbed the 'Coffin Belt' by Patrick Abercrombie though its shape is, perhaps, more reminiscent of an hourglass or dumbell (see Figure 5). Outside this belt, central Scotland, the north-east coast, and the south Wales-Bristol region are the only large areas of high population density.

During the twentieth century improved communications have so reduced the time and cost of transport that it has become increasingly unrealistic to conceive of the industrial geography of the UK as composed of a series of small regions like the north-east coast, or south Wales, or at an even smaller scale, as distinctive manufacturing districts like the Potteries or the Black Country. The traditional association of such areas with specific products (e.g. Lancashire and cotton, Yorkshire and woollen textiles) has gradually been giving way to a more unified and homogeneous pattern. While distinctions remain they are today less important as new locational forces are at work superimposing a new locational pattern upon the existing one. The pattern we observe today is the product of continuing change over a long period of time, a pattern which we must now examine in more detail.

Figure 4
Distribution of population in
the United Kingdom in 1961

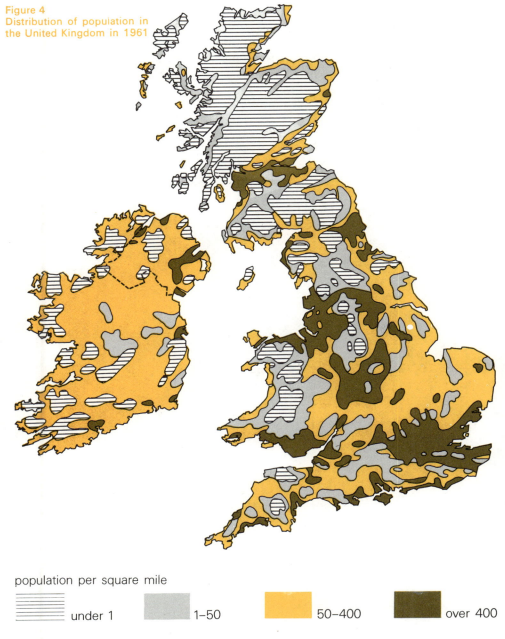

population per square mile

under 1        1–50        50–400        over 400

Source: *Atlas Advanced*, Collins-Longmans Atlases, 1968

## 3.2   The origins of industrial growth

Before the Industrial Revolution Britain's economy was predominantly agricultural. Even in this largely self-sufficient society some specialization had taken place and trade, often over long distances, in such commodities as wool, timber, grain, fish, salt and metals had developed. Population density, though higher in the more fertile agricultural lands of the south, was fairly evenly distributed. Industry, especially the multifarious craft industries, producing for local self-sufficient markets, tended to correspond to the distribution of people. There were exceptions, however, and in these it is possible to perceive the antecedents of those industries which were to become established during the Industrial Revolution. The manufacture of *woollen textiles*, though by the end of the Middle Ages a predominantly rural activity, was centred on East Anglia, and the West Country. *Iron smelting* was another activity that had become increasingly localized, notably in the Weald and the Forest of Dean, though an increasing shortage of fuel (charcoal) encouraged some dispersal in the period before the Industrial Revolution. None of these four rural industrial areas has survived. We can, however, perceive a number of embryonic industrial regions developing on the eve of the Industrial Revolution. The West Midlands, for example, was already by the beginning of the eighteenth century a leading iron manufacturing district, the West Riding had begun to outpace the traditional woollen districts, and Tyneside was the main coal producing and exporting

Figure 5
United Kingdom: Shape of
the industrial axis

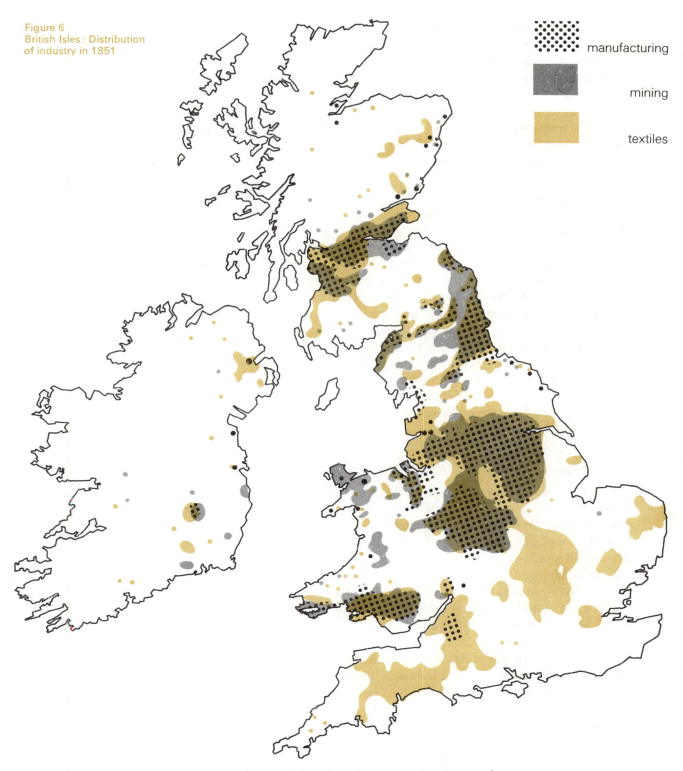

Figure 6
British Isles : Distribution
of industry in 1851

manufacturing

mining

textiles

region. As these and other nascent industrial regions began to develop so the
pattern of generalized, dispersed manufacturing was beginning to yield to one
of regional specialization and concentration. It was the Industrial Revolution
which accelerated this trend until it reached its most complete fulfilment during
the nineteenth century (see Figure 6).

## 3.3   The Industrial Revolution

One of the characteristic features of industrialization is the factory system. The
series of mechanical innovations in the textile and metal industries and the
harnessing of first water and then steam power made necessary the organization

of industry into large units. Since large quantities of bulky fuel (coal) and raw materials were required and since there was considerable weight loss during the manufacturing process industries tended to grow more rapidly on the coalfields in order to minimize transport costs. They may, therefore, be described as *weight-losing*, or *raw material orientated* industries. The expansion and increasing specialization of the developing industrial regions was intensified by improvements in transportation. The creation of the railway network inaugurated a flexible transport system able to handle large consignments over long distances relatively cheaply. Industrial growth was also matched by a rapid rise in the population from 7·4 millions in 1751 to 21·2 millions in 1851. The largest towns, which before the Industrial Revolution were all in southern England, were now, with the single exception of London, located on or near the coalfields. In 1800 no town outside London had more than 100,000 people. By 1851 there were five towns larger than this and by 1891 twenty-three. The great industrial cities experienced spectacular growth.[1] Middlesbrough, the 'Infant Hercules' as Gladstone described it, had not even existed in 1830 when it was decided to extend the Stockton and Darlington railway to the Tees estuary. First as a port and later as an iron and steel town, Middlesbrough's population rose to 7,600 in 1851, 19,000 in 1861 and 40,000 in 1881. It was entirely a creation of the Industrial Revolution and as such was by no means exceptional. The towns grew by absorption as people in the surrounding districts moved to them in search of the greater opportunities they seemed to offer. The migration was usually short distance, a creeping movement over the whole country. By 1851 more people were living in the towns than in the country and the first phase of the Industrial Revolution was complete.

## 3.4    Regional changes in industrial distribution

Our interest in the period since 1850 focuses on the changes in the distribution of industry both between and within the regions already established by the mid-nineteenth century. Basically, these have been (a) the relatively rapid industrial growth of the South and Midlands, and (b) an increasing dispersal of manufacturing. These changes have several causes. First, the development of new, easily transportable, and complementary forms of power such as electricity, gas, and oil have reduced the locational attractiveness of the coalfields. Secondly, increases in technical efficiency have reduced the raw materials required for a given level of output. Thirdly, a whole range of new industries have emerged requiring high inputs of labour and capital, able to achieve internal economies of scale and closely interrelated so that external economies may be achieved. As a result of these factors most industries have been released from their spatial dependence upon natural resources and have become, in a sense, more 'footloose', though this is a term which must be treated with circumspection since all industries are locationally constrained to a greater or lesser degree.

What influence have these changes had on the industrial geography of the UK? On a regional scale the structural changes as a whole have been mirrored by concomitant changes in the balance of regional growth. Unlike the USA locational shifts in the UK may be attributed more to structural characteristics than to differential industrial growth rates (see McCrone, 1969, Ch. VII). Detailed interpretation is bedevilled by the inadequacy of available statistics. We know that the South-east and Midland areas which have specialized in products for which there has been rapidly increasing demand, have enjoyed relatively rapid growth. By contrast the industrial regions which had flourished

---

1   In the decade 1821–31 alone Manchester increased in population by 45 per cent, Leeds by 47 per cent, Bradford by 65.5 per cent, Birmingham by 41.5 per cent, Liverpool by 46 per cent and Sheffield by 40.5 per cent.

during the nineteenth century have experienced relative decline, even stagnation. It is possible to look at this in terms analogous to population movement. Thus the prosperous areas have had a birth rate of new firms which has far exceeded the death rate, whereas in the depressed areas the margin, if it existed at all, was narrow and they had to rely on the inward migration of firms from prosperous regions for their growth (see Law, 1970). The demand for the products in which they have specialized like coal, ships, steel, cotton textiles and so on has been slower in growing than that for the goods of the newer industries such as electronics, chemicals, vehicles, and consumer durables. Slow growth in demand combined with rapid increases in productivity has led to unemployment and even emigration of labour to other regions for there are insufficient alternative occupations.

Colin Clark has attempted to explain the dynamics of the regional pattern by the concept of 'economic potential' (Clark, 1966, 1968 and 1969). He begins by assuming 'regional incomes and distance costs as the most important variables determining the location of manufacturing industries' (Clark *et al.*, 1969, p. 198). Industries are concerned to minimize the costs of their inputs and to have good access to their markets. Increasingly the areas of dense population provide both the inputs (labour, services, information and so on) and are the major markets. As the size of plants increases so location in these areas of greatest accessibility will become more important. This accessibility, or potential, may be measured in terms of (a) regional incomes which are an indication of the level of market demand, and (b) the costs of transporting goods from each potential industrial location to all possible markets.[1] Those areas which have the greatest access to the highest income are the points of highest potential. Conversely areas with sparse populations, with a low income per head, are at the greatest distance from major concentrations of population will have the lowest potentials. Figure 7 shows the potentials in the UK and reflects the hourglass shape referred to earlier. The highest potentials lie in the south-east though if western Europe is included nowhere in the UK falls within the area of maximum potential which is the Ruhr/Low Countries area. Clark concludes that economic potential 'probably explains the mechanism whereby the densely populated areas of the eighteenth century, with a few additions and deletions, became the densely populated area of 1950, (Clark, 1968, p. 299).

While the most obvious manifestations of the 'regional problem' are high rates of unemployment and emigration, there are a number of other symptoms which were identified in an official survey (the 'Hunt Report') investigating the needs of the areas where the rate of economic growth gave 'cause for concern' (D.E.A., 1969). These include indicators of slow growth such as sluggish or falling employment, and a low rate of increase of industrial and commercial premises; slow growth in personal incomes and generally below average earnings; and low female activity rates (i.e. the proportion of women at work). In many of the older industrial areas these economic symptoms are reflected in the dereliction and decay of the physical environment which the Hunt Report submitted 'imposes a significant economic penalty on the area around, since it deters the modern industry which is needed for the revitalization of these areas and helps to stimulate outward migration' (D.E.A., 1969, p. 136). Whatever criteria are used the regions most afflicted are undoubtedly those defined as Development Areas (see Figure 8) to which the government gives assistance in the form of incentives to industry. Additional incentives are offered in the Special Development Areas, those districts most severely affected by the

---

1   There is no space to explain the methods he used in the calculation of potential here. Those who are interested should refer to the articles cited above.

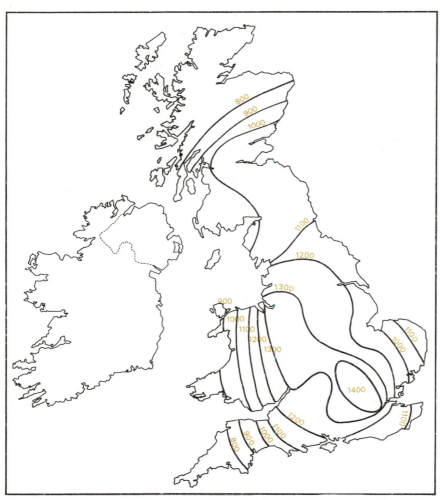

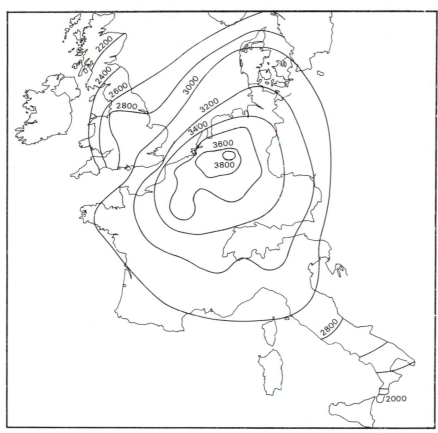

Figure 7
Economic potentials in the
United Kingdom and
Western Europe

For explanation of potential
see text. The potentials
calculated in this Western
Europe example assume an
enlarged Common Market
consisting of the six original
members and Britain, Nor-
way, Denmark, and The
Irish Republic. The values
for both maps are calculated
from 1962 figures

Sources: C. Clark in *Lloyds
Bank Review*, 82, 1966, and
C. Clark and other in
*Regional Studies,* Vol 3,
No. 2, September 1969

Figure 8
United Kingdom: the development and the intermediate areas

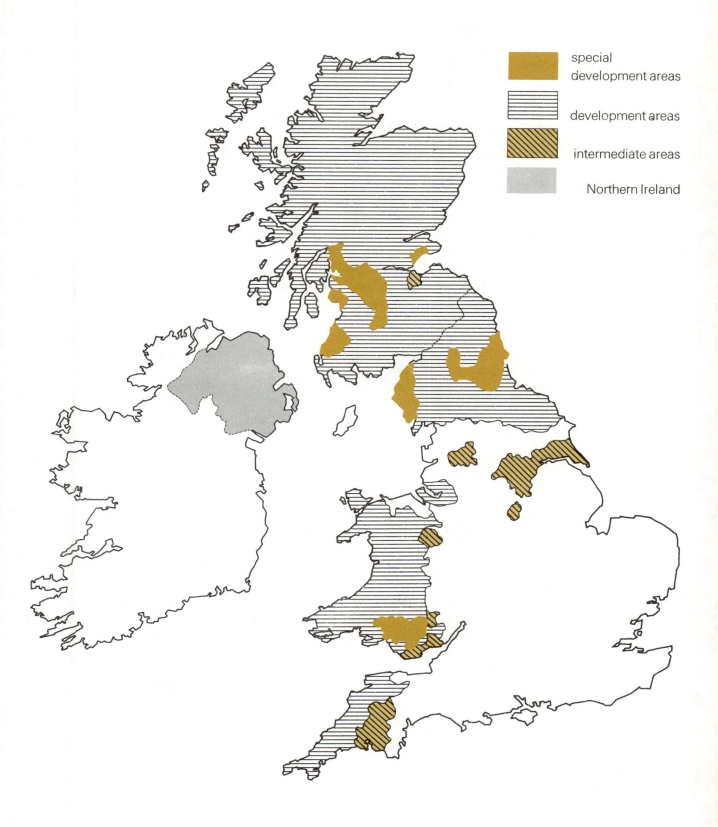

**Source:** *What the Development Areas Offer*, HMSO, 1969

accelerated pit closure programme of the late 1960s. To these a further category of problem region known as the 'intermediate' or 'grey' areas was added in 1969, areas where the problems while not so chronic as those of the Development Areas are sufficient to require government support especially to improve communications and the environment. These areas, geographically intermediate between the Development Areas and the prosperous parts of the country, are also shown in Figure 8. We shall return again to the question of government intervention in the regions and its effect on industrial location in later parts of this block.

## 3.5  Microspatial changes in industrial distribution

In addition to these changes at a regional scale there have also been shifts on a more local scale. As in the USA the process of urban concentration, which resulted in four-fifths of the population living in areas administratively defined as urban by the turn of the century, has to some extent been succeeded by a process of deconcentration. Thus, although the share of the population living in the conurbations of England and Wales (as defined by the Census) has declined (from 38·67 per cent in 1951 to 36·65 in 1961), the surrounding areas have experienced vigorous growth. The outward flow of people has been both voluntary and planned overspill of which the new towns are the most dramatic example. The movement of industry to suburban locations has also proceeded apace. The advent of motor transport has led to greater flexibility in the choice of location and industries have sprung up along arterial roads, on trading estates, and in the country towns within the orbit of major cities. Tertiary activities, especially those like banking, finance, publishing, etc., have proved more stubborn and continue to be concentrated in the central areas of the major cities. This has resulted in the increasing separation of workplace and residence and the consequent long distance commuting that has become such a feature of modern urban life.

## 4  Conclusion

We have seen how the distribution and importance of industry varies in different parts of the world. Industrial specialization and trade have given rise to the emergence of major industrial areas, the most important of which have formed industrial belts. We have considered the pattern as it is expressed in two different economic landscapes, those of the USA and the UK. In both we observed tendencies of industrial concentration and an increasing tendency towards dispersal. This pattern is the result of many individual locational decisions taken over a long period of time. So far, we have concentrated on establishing the background to the more theoretical sections that follow. In these we shall attempt to explain the environment – economic, physical and psychological – in which locational decisions are taken.

The sections which follow introduce some of the theories which have made major contributions to locational analysis. Although the aspects considered relate mainly to the work of this block, location theory is of course relevant to the course as a whole. You should bear this in mind while reading it.

## B   Location theory

## 5   Introduction

The field of location theory is a vast and complex one and we can do no more than indicate here some of its scope and content. In doing so we must be selective both in the theories we examine and in the aspects of theory on which we concentrate. If you wish to pursue the subject in more depth, you will find the theories themselves stimulating but hard going at first. An easier way into them is to begin with commentaries on them such as you will find in your selected reading.[1]

Theories concerned with industrial location have a number of features that account for the location of plants. They tend to be deductive in method and deterministic in approach. They are related to general economic theory and so their arguments rest upon a number of simplifying assumptions which abstract from the real world situation. The most important of these assumptions is that entrepreneurs are motivated by the desire to maximize profit. Although some of the theories, at least, purport to have a general applicability, the values upon which their assumptions rest reflect the capitalist economic system. However, the conclusions which they reach may sometimes be invoked as a justification for state intervention in the location process.

Beyond a broad area of coincidence the various location theories diverge both in the assumptions they make and in the explanations they achieve. These differences are fundamental and are a major reason why, despite the plethora of theoretical writings on location, there is no one generally accepted theory of location. It is possible to classify location theories into three broad groups These are:

(a)   those which emphasize cost factors;

(b)   those which emphasize demand;

(c)   those which are concerned with locational interdependence.

From each of these categories we have chosen a leading theory to illustrate the approach.

## 6   Least cost location theory

### 6.1   Alfred Weber

Weber wrote his *Theory of the Location of Industries* in 1909, and was one of the first to recognize the possibility of a general theory of industrial location [2] In common with other least cost location theorists, he assumed competitive pricing, i.e. a situation where individual firms cannot influence the price of their products which is the same everywhere. At the prevailing price there is unlimited demand and all sellers have unlimited access to the consuming centre It follows from this that the plant which secures the location where lowest costs are incurred will achieve the highest profits.

---

1   Among the commentaries you will find useful are those in the recommended texts by Smith, D. M. (1971) Ch 8; Smith, Taaffe and King (eds.) (1968) pp. 158–172, 186–199; and Karaska and Bramhall (eds.) (1969), pp. 22–41.

2   Von Thünen (see Foundation Course, *Understanding Society* D100 Unit 19, *Rural Land Use*) had written almost a century before on the location of agricultural land use. Wilhelm Laundhardt preceded Weber with an industrial location theory in 1885 but his work was never translated into English, and so has had less impact.

Weber began from the premise that the type of industry is given and that raw materials are dispersed  He considered three groups of cost factors in turn  Two of these he described as general regional location factors. These are (a) transport costs which are expressed in terms of the weight of the raw material or product times the distance carried (and which for theoretical purposes can also subsume the cost of raw materials, fuel and power), and (b) labour costs. The third factor is agglomeration (or deglomeration) which he regarded as a general local factor.

## 6.2 Transport orientation

Weber's theory is best known for its emphasis on transport orientation. Indeed, C. J. Friedrich his translator goes as far as to suggest that 'Weber has succeeded . . . in laying bare the fact that transportation costs are theoretically the most fundamental element determining location' (Weber, 1929, p. xxvi). This aspect of the theory has been covered in the Foundation Course, *Understanding Society* in Unit 20, *Location of Industry*, and in the Reader that accompanies that course, which you might care to refer to at your leisure. Briefly, the minimum transport cost will be at that point where the total costs of moving raw materials and finished products are least. This will, of course, depend on the weight of individual raw materials and the amount of weight lost in processing. Weber illustrated the principle with the simple case of two fixed raw materials and one point of consumption which may be joined to form the 'locational triangle' (see Figure 9). From this he was able to introduce more complex situations of several raw materials, some of which are ubiquitous, some of which lose weight in processing, but all of which can be incorporated to find the minimum transport cost point. He recognized that transport costs per ton/mile may vary according to the mode of transport, the density of traffic, the availability of return loads, the nature of the cargo (perishables attract a premium rate while the rate per ton mile for bulk goods is relatively low), as well as the distance travelled. It is, he suggested, possible to account for all these variables by converting them into relevant units of weight or distance. In attempting to relate his model to reality Weber illustrated the effect of these variations on the locational pattern. For example the cheaper freight rate offered by the canal in Figure 10 will replace the locational triangle originally based on a railway network by another one.

Weber was the first to recognize the concepts of weight loss and transport orientation which helped us to explain the locational pattern of the UK described above. These important concepts had hitherto been neglected in international trade theory which ignored the spatial variable.

## 6.3 Labour orientation

Using his transport orientation model as a framework, Weber went on to explore the importance of differential labour costs as a locational factor. He acknowledged that labour costs may vary according to productivity but was solely concerned with variations that are specifically geographical. Industries will tend to locate at the point of minimum labour costs when 'the savings in the costs of transportation which it involves' (Weber, 1959, p. 103). In figure 11, P is the minimum transport point and is surrounded by a series of *isodapanes*, P is the minimum transport point and is surrounded by a series of *isodapanes*, lines of equal transport costs per ton from P. In the figure isodapanes have been drawn at internals of 5p. Let us suppose there is an alternative location where a saving of 10p per ton may be made in the cost of labour. Clearly, if this location is on the 10p isodapane the saving in labour costs over P would exactly balance the extra transport costs incurred. In this case 10p would be the

Figure 9
Weber's transport orientation

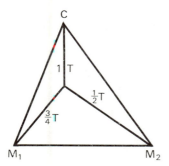

Weber assumed that only weight and distance determined transportation. The optimum location of production will be where the ton mileage of the raw materials and finished products can be minimized. In this example there are two raw materials $M_1$ and $M_2$ which are of $\frac{3}{4}$ and $\frac{1}{2}$ ton respectively and which combine to produce a product of 1 ton consumed at C. Each corner of the triangle exerts a force proportional to the weight attached to it. Thus the optimum location will be nearer to C than $M_1$ or $M_2$ and nearer $M_1$ than $M_2$. Each locational pattern will be specific but the principles applied in this example are general.

Figure 10
Reorientation of transport location

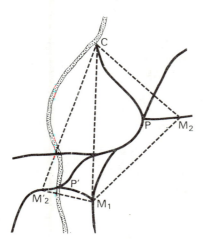

In this diagram $M_1$, $M_2$, $M'_2$ are locations of raw materials, C is the point of consumption, and P, P' are locations of production determined by the principles outlined in Figure 9. If no waterway existed then production would be at P. Although $M_2$ is geographically nearer C than $M'_2$ (the same raw material) the waterway may provide a cheaper means of communication thus making $M'_2$ nearer to C in economic terms. This being so P' would be the optimum location of production.

Source: Figures 9, 10, 12 and 13, A. Weber, *Theory of the Location of Industries*, trans. C. J. Friedrich, University of Chicago Press, 1929

'critical isodapane'. If the minimum labour point lies outside the critical isodapane (e.g. B in Figure 11), the optimum location will remain at P, but if it lies within it (e.g. A in Figure 11), it will pay the plant to locate there since the savings in labour costs more than compensate for the extra transport costs involved in shifting from the point of least transport cost. Such a shift Weber described as a deviation from the minimum transport point. In reality, as Weber recognized, labour orientations were becoming the norm for two reasons. First, developments in transport were increasing the distance over which goods could be moved for a given cost. Put another way, the distance between isodapanes was increasing and therefore more labour locations would be likely to fall within the critical isodapane. Second, the cost of labour involved in producing goods was increasing relative to other costs. Conversely, of course, increasing productivity where less labour was required for a given amount of product would reduce the relative importance of labour locations. We saw earlier how transport improvements and the changing structure of industry go

Figure 11
Weber's labour orientation

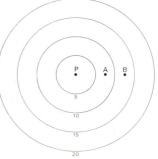

P = minimum transport point
AB = minimum labour point locations

Figure 12
Weber: labour location, reorientation

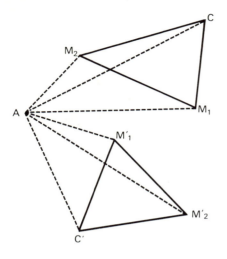

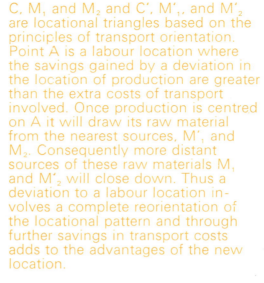

C, $M_1$ and $M_2$ and C', $M'_1$, and $M'_2$ are locational triangles based on the principles of transport orientation. Point A is a labour location where the savings gained by a deviation in the location of production are greater than the extra costs of transport involved. Once production is centred on A it will draw its raw material from the nearest sources, $M'_1$ and $M_2$. Consequently more distant sources of these raw materials $M_1$ and $M'_2$ will close down. Thus a deviation to a labour location involves a complete reorientation of the locational pattern and through further savings in transport costs adds to the advantages of the new location.

Figure 13
Weber: agglomeration locations

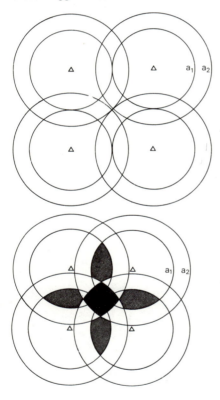

Surrounding the minimum transport cost point of production (the small triangles in the figures) will be sets of critical isodapanes representing the economies that may be achieved by various orders of agglomeration. Thus isodapane $a_2$ while it requires a larger mass of production in order to come into operation will extend further from the minimum point than the lower order isodapane $a_1$. Agglomeration will take place where the critical isodapanes of two or more industrial units overlap. Thus in the top diagram only the higher order isodapanes ($a_2$) overlap and these only for two production units. In the lower diagram the distance between the production units has been reduced. While both sets of isodapanes overlap those of the higher order ($a_2$) overlap for all four production units thereby increasing the potential economies of agglomeration.

far to explain the changing industrial geography of the UK. When a deviation does occur it is likely to result in a total reorientation as new sources of raw materials and markets are substituted for the old (see Figure 12).

## 6.4   Agglomeration

In turning to agglomeration (or deglomeration) his general local factor of location, Weber again built upon the framework he had already established. Once again he used the critical isodapane where the extra costs of transport or labour incurred equals the savings achieved through economies of scale at the point of agglomeration. Again, a total reorientation may result if industry deviates to the agglomeration location. He noted the tendency for such location to increase as declining freight rates increase the spacing of isodapanes and as industrial growth increases the density of manufacturing causing different sets of isodapanes to overlap (see Figure 13). This analysis is not without problems as Weber was well aware. For theoretical purposes transport, labour and agglomeration locations are regarded as fixed and separate points. In reality this will rarely be the case. Although, as defined by Weber labour locations were geographically separate from points of agglomeration, he recognized that labour locations are, by their very nature, also points of agglomeration. Such 'accidental agglomerations' as he called them, would reinforce the tendency to deviate from the minimum transport point. However, increasing size of agglomeration might eventually bring increasing costs and deglomeration (or decentralization of industry) might result. Increases in the cost of land and other factors have caused some deconcentration of industry as we have seen from our examples of the USA and the UK.

The least cost approach to location theory, as embodied in Weber's work, has obvious limitations. In concentrating on the plant he failed to explain the aggregative patterns, an understanding of which is so important for regional economic planning. His theory focused on an optimum substitution of cost factors. The most important criticism is that, by holding demand constant and effectively at a single point, he prevented it from playing any actively determining role in plant locations. His theory, therefore, represents a special case and has no universal applicability. Weber was aware of many of the variables which he was unable to incorporate explicitly into his theory. Despite its failings his was a pioneer work and has remained a formative influence on that of his successors.

## 7   Market area theory : August Lösch

Market area analysis attempts to account for the dimension that Weber omitted – the influence of demand upon location. The basic idea is that since buyers are scattered over the land surface it follows that each plant will be able to control a territory within which its products will undersell those of less favourably located rivals. Within this 'market area' the plant has, in theory, a spatial monopoly. The primary concern of this approach is to explain the size and shape of market areas.

The outstanding market area theorist was Lösch (1954). Lösch's theory[1] is ambitious for it represents the first attempt to build a comprehensive general location theory. He began with a survey of the problem and proceeded first to an analysis at the level of the individual firm and then at the regional scale. Into this simple, static model he then introduced the inequalities of the real world situation and his concluding sections illustrated his theory by reference to a range of examples of market areas drawn from Europe and America.

We are here only concerned with what he had to say about the location of the firm. Unlike Weber, Lösch did not assume production and consumption are at fixed points or that demand is given. He exposed one serious defect of Weber's theory, namely his insistence on least cost locational motivation (when Weber was writing most industries probably were least cost orientated). To concentrate on costs alone 'is as absurd as to consider the point of largest sales

---

[1] Lösch's theory is contained in *The Economics of Location* (trans. 1954). There is also a short paper by Valavaniss, S. in Smith, Taaffe and King (eds.) (1968), pp. 69–74.

as the proper location. Every such one-sided orientation is wrong. Only search for the place of greatest profit is right' (Lösch, 1954, p. 27). Lösch thus explicitly excluded non-economic factors from his analysis.

He commenced his analysis by assuming an *isotropic plain*, that is a homogeneous land surface with an evenly distributed population of self-sufficient farm households each having the same tastes and similar technical capabilities, in short, a surface from which all irregularities and non-economic factors have been abstracted. Incidentally, the consequence of this assumption is that Lösch failed to incorporate production cost differentials into his theory. He further assumed identical production, identical demand curves for each buyer of each product and that transport costs are proportional to distance. In such a situation the size and shape of the market area will be a function of the price of the product and the freight rate. On this ideal surface one farmer decides to produce beer over and above its own needs. AQN in figure 14a is the demand curve for beer (all households having the same demand) and MP is the price of beer at the brewery. Prices above MP represent the price at the brewery plus the cost of transport. Therefore at P the demand will be PQ while as distance from P increases and price rises so demand will fall until there is none (A in Figure 14a). The Line PA may, therefore, represent the distance from the production point P to the limit of demand A.

If we project this simple economic graph onto our plain and rotate it around the point P we shall produce a three dimensional demand cone showing the total volume of sales from brewery P (Figure 14b). In this demand cone the line PA represents the limit of demand and rotated about P describes a circle which we can call the *market area* of the brewery.

Other farms may find that demand for beer is sufficient to outweigh the costs of producing beer surplus to their own requirements. Transport costs and the economies gained by increasing the scale of production will be the determining factors. New producers will continue to enter the market until all abnormal or excess profits are eliminated, i.e. marginal costs equal marginal revenues, and the market is saturated. In this way a network of market areas covering the plain will be built up and a point of equilibrium reached. Assuming that buyers will purchase from the nearest producer we would expect the shape of market areas to be circular as this is the shape that has the smallest circumference for the area it contains (i.e. the average distance to consumers

Lösch: derivation of a demand cone

Figure 14a

Figure 14b

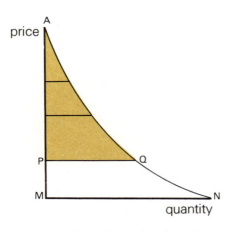

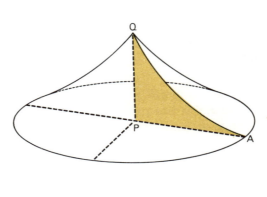

Source: A. Lösch, *The Economics of Location*, trans. W. F. Stolper, Yale University Press, 1954

within it is smaller than for any other geometric shape of market area). But circles can only cover a surface if they overlap and if they do not cover it gaps will be formed. Hexagons eliminate this problem and are, in fact, the idealized shape adopted by market areas.

Each product will have a different size of market area depending on the relative importance of transport costs as opposed to economies of scale. Thus there will be a network of market areas for each product. Together these market areas will produce a system of networks which comprise an economic region or landscape.[1] By rotating these networks around a common producing centre Lösch demonstrated that there will be sectors containing production centres which concentrate a wide range of activities. Conversely, there are other sectors where production will be more dispersed.

By relaxing certain of his assumptions Lösch was able to confront his idealized model with the real world situation. Price changes are of obvious importance for they will cause an expansion or contraction in the number of consumers served. The actual amount of new custom attracted will depend on the elasticity of demand but we would normally expect a price fall to extend the area of the market and to increase the intensity of demand within it. Differential freight rates and spatial pricing strategies (see also Unit 8 of this block) may lead to some distortion of the idealized pattern. Similarly variations in topography, population density, accessibility, entrepreneurial ability, and cultural and political attitudes must be accounted for if theory is to be tested against reality. This reality may appear chaotic but it is 'the chaos which is really nothing but order in disguise' (Lösch, 1954, p. 220). Lösch demonstrated the underlying order in locational patterns by assuming rational economic man. In fact as he acknowledged entrepreneurial choice 'rests upon subjective considerations. [The entrepreneur] will of course bear objective facts in mind, but these alone cannot dictate locations' (Lösch, 1954, p. 16). In his theory, however, Lösch neglected the behavioural aspects of locational choice.

The market area approach has certain limitations. The highly abstract economic landscapes are composed of industrial concentrations where a number of activities are concentrated but in which individual activities are dispersed, each plant controlling its own market area. Lösch ignored those situations where competitors locate at the same point. In taking into account the influence of competitors upon each other we come to our third theoretical approach, that of locational interdependence.

## 8   Locational interdependence : Harold Hotelling

The locational interdependence approach, like that of market area analysis, is concerned with the impact of demand upon location. It ignores cost factors altogether and therefore cannot be applied as a general case. However, its emphasis on the interaction of entrepreneurial decisions introduces an element of locational behaviour we have not looked at so far.

The best known model of locational interdependence is that propounded by Harold Hotelling in his paper 'Stability in competition' (1929). He assumed an evenly distributed population along a linear market served by two competing entrepreneurs (a situation known to economists as duopoly) each with equal production costs and capable of supplying the entire market, producing two identical products. Demand is assumed to be infinitely inelastic[2], the entrepreneurs may relocate without cost, costs of production are assumed equal

---

[1]   Those familiar with the Foundation Course, *Understanding Society*, D100 Unit 22 *The Size and Spacing of Settlements*, will recognize the relevance of Lösch's work to central place theory.

[2]   The concept of inelasticity is of fundamental importance in economic theory, i.e. demand is not responsive to changes in price. It is introduced in the Foundation Course, *Understanding Society*, D100 Units 10–12, *Economy and Society*.

everywhere, the only variable cost being transport, which varies directly with distance. This model is usually illustrated in terms of two ice-cream salesmen (A and B) on a beach (figure 15). Each has a monopoly over a certain market area. The boundary between their market areas is determined by the point of indifference (i.e. equal price) between them.[1] In order to compete for more customers they may alter their prices. Hotelling argued that prices will find an equilibrium level at which both entrepreneurs can maximize profits. In order to enjoy excess profits they would either have to reach a tacit agreement (always likely to be broken) or indulge in a price war in which one is eliminated. This might be both costly and difficult especially when demand for the product is infinitely inelastic, that is, when it does not increase as price falls. Unable to compete through price they may try to compete by changing their location. If A were to move just to the left of B he would increase considerably his market share at the expense of B. B might retaliate by moving to the left of A and so on until both find themselves clustered at the geographical centre of the market (Figure 15b). Once again a situation of stable equilibrium has been reached.[2] However, this is a costly situation for it results in excessive transport costs for half the customers in our linear market. Indeed, if we relax the assumption of infinite inelasticity, both A and B may sacrifice some distant customers in moving especially if demand for the product is elastic, that is, sensitive to price movements. They may overcome this to some extent by adopting spatial price discrimination whereby the price to nearby customers is raised, while it is reduced for those further away. In doing so some absorption of transport costs would be necessary. Such a policy is likely to succeed when prices are lowered in competitive markets and increased for those customers who are nearby within the area over which the seller has a market monopoly. The degree to which prices are raised or lowered will, of course, be a function of the price responsiveness or elasticity of demand for the product. If such a policy were to be adopted prices would not increase directly with distance but would tend to become more uniform. Pricing strategies and their influence on location will be discussed in more detail in Unit 8.

The optimal solution in the case of duopoly discussed here would be when the market was evenly shared and each seller was at the mid-point of his market (known as the quartile points of the line, see Figure 15c). Here transport costs for customers are minimized and sales for sellers are maximized. This reinforces a point already made that market forces need not produce optimal solutions and that, therefore, some intervention (or planning) may be justifiable.[3]

In reality the situation is much more complex. Other possibilities are introduced if we relax some of Hotelling's other assumptions. If there are more than

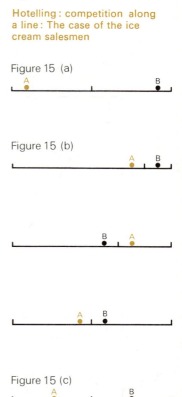

**Figure 15**

Hotelling: competition along a line: The case of the ice cream salesmen

Figure 15 (a)

Figure 15 (b)

Figure 15 (c)

---

[1]   This may be expressed in the following equation:

$P_1 + cx = P_2 + cy.$

where $P_1$ is the price of ice-cream at A

$P_2$ is the price of ice-cream at B

$c$ is the cost of movement per unit of distance and,

$x, y$ are distances respectively from A and B to the boundary between their market areas.

[2]   Several writers, however, argue that instability is more likely since there will be considerable price fluctuation as sellers approach each other. This will be especially so when there are more than two sellers in the market. Each seller, by assuming the others will remain where they are is likely to move to a more profitable location. This may injure one of his rivals who then decides to move and so on.

[3]   From what has been said it is clear that several variables must be considered when we are attempting to generalize about the locational strategy of competing sellers. Among the most important of these are (i) the elasticity of demand for the product. Where it is infinitely inelastic then localization is likely to result as sellers do not lose their more distant customers while at the same time locating at the centre of the market, thus preventing their rivals from gaining a locational advantage. Conversely, dispersal is likely with highly elastic demand as sellers seek to reduce the transport costs of their more distance customers; (ii) transport costs, where these are high dispersal is likely, and vice-versa.

**Figure 16**

Isard: modified Losch system
In this presentation only three sets of commodities are shown and the hierarchy of central places has been simplified to avoid confusion. The straight lines radiating from each centre are transport routes. Away from the centres and the transport routes population is sparse and mainly agricultural with a thin scatter of service centres. The zonal pattern of agricultural land use is indicated by the shaded bands in the bottom right hand corner

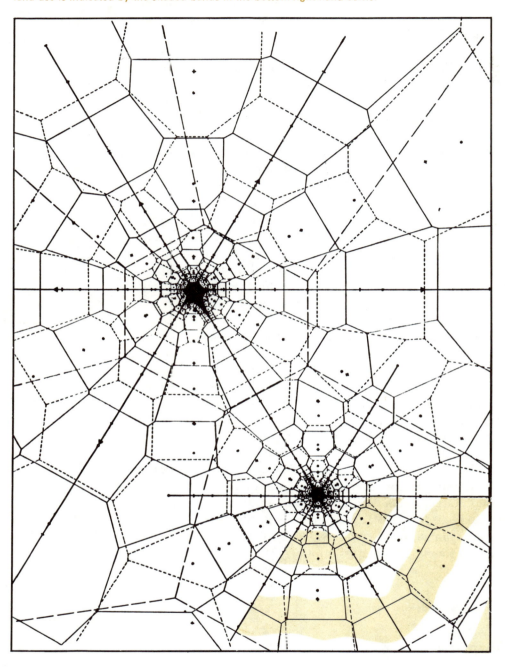

market area boundaries

Source: W. Isard *Location and Space-Economy*, MIT Press, 1956

two sellers then dispersal is likely to result as each seller seeks to control part of the market. Conversely, if we do not assume an even distribution of population then concentration is likely to be encouraged. Population clusters are both cause and effect of the concentration of economic activities. Population centres are points of maximum accessibility where buyers can minimize the time and distance of travel. There they may choose from a variety of goods. As producers agglomerate at the population centre so there is increasing product differentiation and economies of scale both internal and external may be achieved. The market continues to expand and attracts new producers until a point is achieved where new firms can no longer be accommodated either owing to lack of room at points of maximum accessibility or because the barriers to entry (the minimum financial outlay needed to enter the market) are too high. However, the competition for sites causes rents to be bid up and thus part of the profit earned at the site is converted into rent paid to the landholders. Thus some deconcentration of activities is encouraged and new clusters arise in other parts of the market – a situation of dispersed concentrations.

Each of these three groups of theories offers a particular insight into industrial locational patterns. Each of them also fails to incorporate important parameters that help to determine these patterns. We finish this section by looking briefly at two attempts to integrate the various theoretical approaches

## 9   Integrative approaches : Walter Isard and Melvin Greenhut

Walter Isard attempted to consolidate the various location theories and to provide synthesis 'embracing the total spatial array of economic activities' (Isard, 1956, p. 57). To do this he treated all aspects of location – the plant, urban and agricultural location – at all scales – local, regional and international. Much of his analysis was devoted to a restatement and refinement of existing theories especially those of Weber and Lösch, and having ranged over the whole area of location theory, Isard concluded with his attempted synthesis. The Weberian approach emphasized the pull of specific localized raw materials and ignored the spatial nature of consumption. Lösch's market areas were spatial constructions but he assumed evenly distributed consumption, a situation that would be impossible once industrial concentration had taken place. Isard transformed the neat even-sized hexagonal market pattern that Lösch evolved into one that is at once more complex and distorted (see Figure 16), but which conforms more to reality. He showed that the density of demand will be greatest near the major concentrations and therefore the market area for each product will be smaller there and will gradually increase in size as distance from the centre increases, production becomes more extensive, and population more sparsely distributed. Isard then superimposed his modified Löschian landscape onto a concentration of heavy, localized raw material using industry such as Weber's scheme might produce. He then transposed part of this model into a commodity flow pattern in order to integrate the local (micro) scale with the regional and international (macro) scale. Some idea of his scheme may be gained from figure 17. It was his ambitious attempt to place an all inclusive location theory within the framework of economic theory rather than any startling new insights that is the important feature of Isard's work.

Melvin Greenhut devoted most of his book to a critique and analysis of the various schools of location theory. His contribution is important in two respects. First, as the title of his book, *Plant Location in Theory and in Practice* suggests, he stressed the importance of testing the relevance of theoretical assumptions by empirical investigation. Secondly, he underlined the importance of non-economic factors in industrial location.

Location theories of the type we have been looking at are deductive. There is therefore a need to test their validity by examining the locational patterns adopted by specific industries. Greenhut did this and demonstrated that the deviation of the real from the theoretical may be considerable and may be explained by a whole series of variables not accounted for in the models, such as, for example, freight rate structures. He also argued that since existing theory has been predicated upon large firms there has been a tendency to overlook the purely personal factors, such as business contacts that are of prime importance to the small firm. He concluded by arguing that 'it is by focusing attention on all of these factors that insight into the underlying reality of plant location (spatial economics) is gained' (Greenhut, 1956, p. 291).

Figure 17

Isard : Industrial location. Urban, regional and international relationships

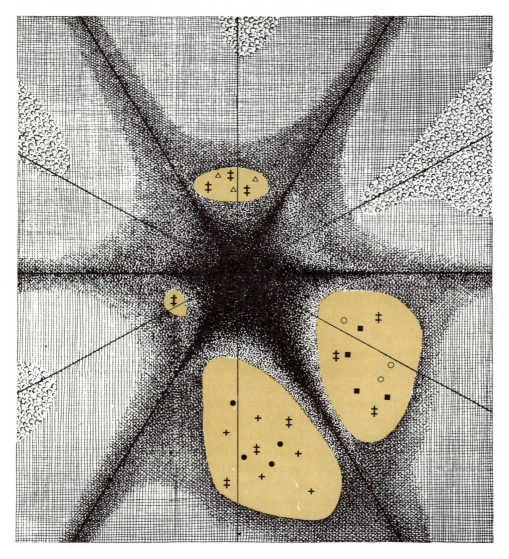

○ ＋ ■ ● △   firms manufacturing specific commodities
from localised raw materials and concentrated
at the urban centre to achieve internal
and external economies of scale

‡   firms manufacturing miscellaneous
items or using ubiquitous raw materials

In this series of diagrams we move from the micro scale of the city to the international scale. (a)   The introduction of firms using localized raw materials acts as a powerful stimulus to the growth of the city. Around the city centre congregate commercial and service activities in a dense cluster shown by the small black dots. Outside this core are the industrial areas within which are four distinct industrial districts each tending to specialize in order to achieve economies of scale. Beyond these areas are the residential districts and open space.

Source : Figures 17 a, b, c, d, W. Isard, *Location and Space-Economy*, MIT Press, 1956

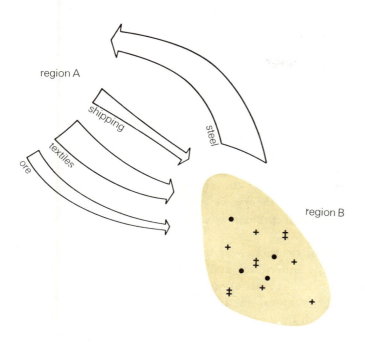

(b)   From the static pattern of urban land use we introduce dynamic relationships. Various inward and outward flows of people, services, and goods will be generated. Here, at the regional scale the lower central industrial district of the previous diagram is shown as region B and the firms marked ● within it are exporting steel to region A. Region A which has an absolute advantage in the production of ore, textiles and shipping, exports these commodities to region B. The arrows are proportionate in width to the value of the commodity.

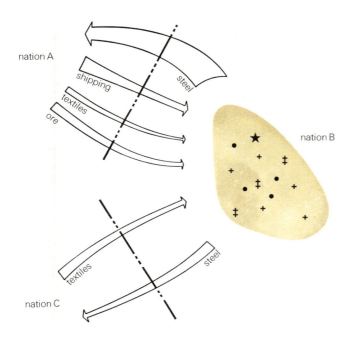

(c)   At the international scale where the factors of production are relatively immobile trade is on the basis of comparative advantage. Nation C although at an absolute disadvantage in the production of each commodity is able to specialize and trade on the basis of comparative advantage in textiles. The magnitude of flows is altered and nation B unable to acquire all the textiles it requires develops its own textile industry (marked ★ ).

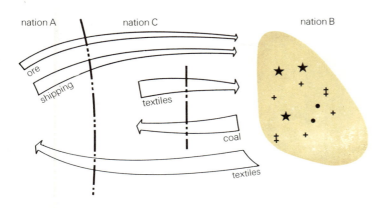

(d)   If the relative geographical position of the three nations is altered then the magnitude of the flows and the industrial structure of each nation will undergo transformation. As the diagram shows nation B's structure is quite different from that obtaining in the previous situations.

## 10   Conclusion

In his emphasis on personal factors Greenhut was anticipating those who approach locational questions from a behavioural viewpoint. Observation of the real world induces an awareness of the idiosyncrasies that are contained within any locational pattern. We can find numerous examples of plants that survive, indeed thrive, in locations which theory would predict as sub-optimal if not hopelessly uneconomic. Clearly the observed pattern cannot be explained fully in terms of economic theory, still less in terms of theories which are based upon such simplifying assumptions as constant demand (least cost theories) or an isotropic plain (maximum profit theories). Instead of the omniscient, rational, profit seeking or cost minimizing individual of theory we must consider the ill-informed, irrational, and perhaps wilful entrepreneur who actually takes locational decisions.

That subjective factors play a part in locational decisions was not denied by the classical theorists. Weber remarked on the individual and arbitrary nature of many decisions and Lösch declared that choice 'rests upon subjective considerations' (Lösch, 1954). Important as behavioural factors are they operate within a broadly economic framework which they may distort but not destroy. Economic forces will determine the limits (spatial margins) within which a location must be sought. The size, efficiency and location of plants will vary considerably within the limits imposed by economic viability. Some plants will be eliminated in the process of trial and error that imperfect knowledge of all the relevant factors implies. Those which survive will do so in a variety of sub-optimal locations which reflect the various attitudes and motives of the decision takers.

What are the subjective factors which influence locational behaviour? What is their relative importance as against non-economic factors? These are important questions and the answers will vary according to the type, size and structure of the industry we are considering. Indeed it may be impossible to give accurate answers to such questions. It is, however, to questions such as these that we turn in the remainder of this block.

# References

BERRY, B. J. L. (1970) 'The United States in the Year 2000' in *Transactions of the Institute of British Geographers*, No. 51, Nov. 1970, pp. 21–53.

CHISHOLM, M. (1970) *Geography and Economics*, (2nd ed.) London, G. Bell & Sons.

CHORLEY, R. J. and HAGGETT, P. (eds.) (1968) *Socio-economic models in Geography*, London, Methuen.

CLARK, C. (1966) 'Industrial Location and Economic Potential' in *Lloyds Bank Review*, 82, pp. 1–17.

CLARK, C. (1968) *Population Growth and Land Use*, London, Macmillan.

CLARK, C., WILSON, F. and BRADLEY, J. (1969) 'Industrial Location and Economic Potential in Western Europe' in *Regional Studies*, Vol. 3, No. 2, Sept. 1969, pp. 197–212.

D.E.A. (1969) *The Intermediate Areas*, Cmnd. 3998, London, HMSO. (The Hunt Report.)

FUCHS, V. R. (1962a) *Changes in the Location of Manufacturing in the USA since 1929*, New Haven, Yale University Press.

FUCHS, V. R. (1962b) 'Determinants of the Redistribution of Manufacturing in the US since 1929' in *Review of Economics and Statistics*, Vol. XLIV, No. 2, pp. 167–77.

GREENHUT, M. L. (1956) *Plant Location in Theory and in Practice*, University of North Carolina Press.

HOTELLING, H. (1929) 'Stability in Competition' in *Economic Journal*, Vol. 39, pp. 41–57.

ISARD, W. (1956) *Location and Space Economy*, New York, Technology Press, John Wiley.

ISARD, W. (1960) *Methods of Regional Analysis*, Cambridge, Massachusetts, MIT Press.

KARASKA, G. J. and BRAMHALL, D. F. (eds.) (1969) *Locational Analysis for Manufacturing*, Cambridge, Massachusetts, MIT Press.

LAW, C. M. (1970) 'Employment Growth and Regional Policy in North West England' in *Regional Studies*, Vol. 4, No. 3, Oct. 1970, pp. 359–366.

LÖSCH, A. (1954) *The Economics of Location*, (trans. W. F. Stolper) Yale University Press.

MCCRONE, G. R. (1969) *Regional Policy in Britain*, London, Allen and Unwin.

ROSTOW, W. W. (1967) *The Stages of Economic Growth*, Cambridge, Cambridge University Press.

SMITH, D. M. (1971) *Industrial Location: An Economic Geographical Analysis*, New York and Toronto, J. Wiley & Sons.

SMITH, R. H. T., TAAFFE, E. J. and KING, L. J. (eds.) (1968) *Readings in Economic Geography: the Location of Economic Activity*, Chicago, Rand McNally.

SMITH  W. (1968) *An Historical Introduction to the Economic Geography of Great Britain*, London, G. Bell & Sons.

SPATE, O. H. K. and LEARMONTH, A. T. A. (1967) *India and Pakistan*, London, Methuen.

WEBER, A. (1929) *Theory of the Location of Industries*, (trans. C. J. Friedrich), University of Chicago, (originally published in German in 1909).

ZELINSKY, W. (1962) 'Has American Industry been Decentralising? The Evidence for the 1939–1954 Period' in *Economic Geography*, Vol. 38, July 1962, pp. 251–68.

# Unit 4
# External costs

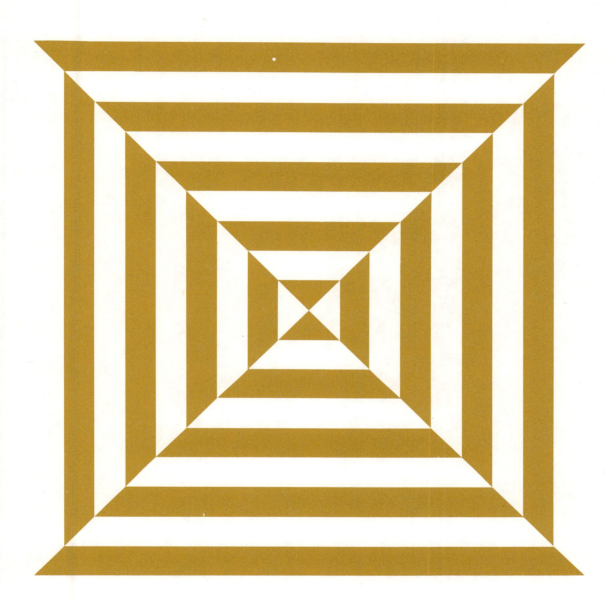

prepared by Geoffrey Edge
for the course team

# Unit 4 Contents

# External costs

## 1  Introduction

The aggregate pattern of industrial location whether on a world or on a regional scale is made up of individual factories and firms. In order to understand the pattern described in Unit 3 explanations have to be sought about the location decisions, operating costs and demand conditions of the individual plant. This part of the block is concerned with the way in which the factors of production which each plant uses differ in cost from one location to another and examines why those differences occur. Most plants require labour, capital, power and raw materials in order to carry out their productive operations. They also need transport to bring in their 'raw materials' and carry their finished products to the market. Land is needed for factory buildings and storage while managerial expertise, or enterprise, is needed to run the factory. Naturally the quantities of these inputs demanded will differ from plant to plant and from industry to industry. In addition their costs to each plant will also differ. In some cases the cost of a single production input will be so great in relation to all other factor costs that the operating costs of firms in an industry can be closely estimated merely by mapping the variation in cost of that factor. More normally no single factor input provides the dominant production cost in an industry in which case actual operating costs can only be estimated by mapping the combined costs of a wide range of factors. This task is of course a difficult and lengthy one and has only been attempted in a limited number of cases. Cost variations in labour, enterprise, capital, power and raw materials will be discussed first of all, and followed by an examination of transport and land costs. Empirical examples of the actual pattern of cost variation will be cited wherever possible.

## 2  Labour costs

Any consideration of labour costs involves two major elements:
(a)  How labour costs vary from place to place – a static element.
(b)  What forces help to equalize labour costs – a dynamic element.

### 2.1  Labour cost variation from location to location

Differences in labour cost occur on almost any spatial scale from the very large to the very small. On the world scale we can recognize great areas such as India and Africa where labour is relatively cheap and plentiful while there are others such as North America and western Europe where it is expensive and in strong demand. In addition to these continental differences there are differences within countries. There is a literature going back over twenty-five years on the causes of low wages in the southern USA as compared with wage levels in the Manufacturing Belt of the north-east. On a smaller scale there are differences in wage levels and labour costs in the UK with costs tending to be higher in the prosperous regions of the Midlands and the South-East, and lower in the north of England and Scotland. There is also evidence to suggest that wage levels differ not only from town to town but also from factory to factory. This suggests that labour costs are likely to be important no matter on what geographical scale the variations in the costs of the factors of production are being considered.

## 2.2   The size of the labour force

One of the most important determinants of the cost of labour is the supply of labour in relation to the demand for it. The size of the population in a particular location has a major impact on supply levels as does the proportion of the population who are economically active. It is quite possible to get two areas which have the same population, but which have different activity rates. At a national level the age of leaving school and of retirement are important parameters determining the size of the labour force. If the age of school leaving is raised or the age or retirement lowered then the number of people available for employment at a particular location will fall and assuming that demand remains constant then wage levels, and hence labour costs, will rise.

The activity rate of females is often of key importance in developed countries where the potential for attracting new men into the labour force is strictly limited. In Britain in 1956, 52.6 per cent of women were not employed and this figure ranged from a low of 49.0 per cent in the North-West, where cotton textiles has long provided a source of female employment, to a high of 61.3 per cent in Wales where the opportunity for women to work is not very great. Detailed differences between regions in the activity rate of women can of course be explained largely in terms of industrial structure, but a considerable residue remains which presumably reflects the British social tradition that married women should not work. This contrasts sharply with the situation in East Germany where many women go out to work. This is because communal child care facilities are extensive, compared to Britain. Women's wages are also required for household expenditure on some non-essential goods.

The level of female employment is also important in countries like the UK and the USA as a factor directly affecting labour costs. Because female wages are lower than male in many occupations, an industrialist can minimize his wages bill by employing female labour, wherever possible, in preference to men and by locating in areas where there is a large pool of female labour. V. R. Fuchs (1967) for example found that in non-agricultural activities in the USA in 1959 white male wages ranged between $2.54 and $3.09 per hour while white female wages ranged between $1.56 and $1.97 an hour. These are average figures but they do indicate that women's wage levels were around a third less than men's wages. Clearly an industrialist anxious to reduce costs would substitute female labour for male labour wherever possible. Such disparities also exist in Britain although legislation has now been passed to equalize wage levels and thus remove the opportunity for British firms to substitute cheap female labour for more expensive male labour.

## 2.3   The degree of unionization

Another factor which influences the spatial variation in wage levels is the degree of unionization of the labour force. In Britain unions tend to negotiate wage agreements on a national basis. These set a minimum level below which wages cannot fall. Wages do however vary very considerably above the minimum with much detailed wage bargaining being carried out at a plant level. In those factories where union membership is large and active, wage levels can often be forced up way above the basic rate. Preliminary investigations in Britain certainly suggest that wage levels and hence wage costs can differ quite markedly even between factories in the same town, so that any industrialist seeking to minimize his labour costs could gain by attempting to identify areas where unions were relatively weak. In addition the relative lack of unionization of female labour means that low wages can be paid and costs reduced in areas where women workers are plentiful.

In the USA the effect of unionization on wage levels and wage costs has received much greater attention than in the UK. There the degree of unionization not only differs much more than in the UK but the degree of legal protection afforded to trade unions and to workers in general also differs from state to state. Table 1 taken from the work of De Vyver, compares the number of union members in the southern USA with those employed nationally in selected industries. Notice that with the sole exception of the tobacco industry union membership in the south is much lower than one would expect to find if the distribution of union members followed the same pattern as the distribution of employees. The lumber and timber industry is perhaps the most striking example where 54 per cent of national employment in that industry is concentrated in the southern USA but only 13 per cent of the union membership is there.

Table 1

Comparison by industry of percentage of total national employment and percentage of total union membership in selected southern states[1] of the USA.

| industry | total employment | employment in South | percent of total employment in South | percent of total union membership in South | estimated total union membership for industry[c] | estimated union membership in South[d] |
|---|---|---|---|---|---|---|
| tobacco manufactures | 103,289[a] | 66,057[2a] | 64 | 75 | 45,571 | 34,180 |
| textile mill products | 1,147,194[a] | 544,521[3a] | 47 | 20 | 460,770 | 103,967 |
| apparel and related products | 972,897[a] | 120,833[a] | 12 | 3 | 798,010 | 25,968 |
| lumber and timber basic products | 388,665[a] | 208,515[a] | 54 | 13 | 90,000 | 11,800 |
| furniture and fixtures | 282,780[a] | 68,304[a] | 24 | 10 | 95,000 | 9,635 |
| construction industry | 1,073,655[b] | 178,216[b] | 17 | 9 | 1,428,000 | 135,363 |
| coal industry | 370,636[b] | 90,764[b] | 24 | 15 | 600,000 | 87,500 |

1 Alabama, Arkansas, Florida, Georgia, Kentucky, Louisiana, Mississippi, North Carolina, South Carolina, Tennessee, Virginia

2 Excludes Alabama, Arkansas and Georgia for which employment figures are not available

3 Excludes Arkansas

Sources:

a U.S. Department of Commerce, Bureau of the Census, *Census of Manufacturers 1947*

b U.S. Department of Commerce, Bureau of the Census, *Sixteenth Census of the United States*

c U.S. Department of Labor, Bureau of Labor Statistics, Bulletin No. 980

d F. T. de Vyver, 'The Present Status of Labor Unions in the South – 1948', *Southern Economic Journal*, XVI (July 1949)

Source: Vyver, F. T. de (1951) 'Labor factors in the industrial development of the South' in *Southern Economic Journal*, Vol. 18, pp. 189–205.

Low trade union membership does not necessarily mean low wages although it seems to be generally agreed that the lower wage levels of the southern USA are at least partly explained by the absence of union pressure for increases in pay and fringe benefits.

The second aspect of union activity on labour costs in the USA is provided by labour legislation. The details of labour legislation differs from state to state but generally the levels and duration of both unemployment and accident benefits paid to workers in the southern USA are less than in the Manufacturing Belt. In addition a number of southern states have passed anti-union legislation outlawing the closed shop and in some cases imposing severe restrictions on the right to strike. Finally in the southern states legislation controlling hours of work and safety conditions tends to be much less stringent than it is in the north. Because of this a manufacturing concern seeking to hold down its labour costs and have the maximum degree of freedom in the use of its labour force will tend to be attracted to the southern USA.

## 2.4   Variations in productivity

So far in our discussion of the impact of labour costs on the costs of production at differing locations we have tended to equate wage levels with labour costs. While clearly there is some relationship between the two, the exact impact of a given level of wages on total labour costs will be determined by the productivity of the work force. Naturally a highly paid but highly productive workforce may produce lower labour costs than a lower paid but much less efficient force. Productivity levels too seem to vary from one geographical location to another so that in theory an industrialist can minimize his labour costs by choosing to locate his factory in areas where the productivity of the labour force is high. In practice, however, the measurement and interpretation of geographical variations in productivity is highly problematic.

One of the earliest attempts to measure productivity variation in the UK was made by H. W. Singer and C. E. V. Leser (1948). They compared differences in productivity for England and Scotland based upon figures for output per worker as recorded in the 1935 Census. They found that productivity seemed to be 5 per cent to 10 per cent higher in England and Wales than it was in Scotland. In the discussion which followed the publication of their results several criticisms were levelled at their methods of analysis and in 1950 C. E. V. Leser tried to refine his analysis and to extend it to measure the variations in productivity levels of the labour force in each region of the UK. Ever since 1950 attempts have been periodically made to measure productivity variations in Britain. An example of more recent work is that carried out by P. E. Hart and A. I. Macbean in 1961.

The major criticism of all this work is that many of its conclusions are highly tentative. It seems fairly certain that variations in productivity of the labour force from one region of Britain to another have relatively little to do with the fundamental energy, efficiency or dedication to work of different groups of employees. In fact they can largely be explained by the different degrees of importance of particular industries in the employment structure of one region as opposed to another. Some industries are more highly capital intensive than others and in these productivity per worker is consequently high. In those regions where capital intensive industries are important employers of labour the average productivity level for the whole region is high. As a result our original vision of an employer searching the country for areas where labour is highly productive fades and we are left with the conclusion that there is little evidence to suggest that the productivity of the labour force will differ greatly if a factory is located in Central Scotland or in the West Midlands. Such evidence might be produced by comparing the productivity of labour in the same industry from region to region, but as yet this has not been done.

The above analysis does not suggest that productivity levels do not change from one factory to another but it does suggest that, as far as we know, they do not change in any systematic geographical way. However it implies that since productivity levels in a particular industry do not, as far as we know, differ substantially then the mapping of variations of WAGES in that industry will provide a general approximation of variations in labour COSTS from one geographic location to another.

## 2.5   The impact of demand

Although the size of the labour force, the degree of unionization and possible variations in productivity all have to be examined to assess their impact on labour costs, the major determinant of wage levels and hence costs is the level of demand for labour. This will differ from area to area according to the changing mix of industries and the level of demand for their products. In areas where declining or slowly growing industries predominate, the demand for labour is likely to be either static or declining. In those circumstances higher than average unemployment levels will develop as the size of the available labour force increases with population growth. Much of the north and west of Britain has had unemployment levels one or two per cent above the national average ever since 1945. In contrast to the regions where demand for labour is low, there are those dominated by rapidly growing industries where demand is constantly rising. A whole series of articles have been published on the variations in the demand for labour at the regional level and the way in which these relate to the industrial structure. For example Leser (1949) has written on changes in employment levels in the regions of Great Britain during the period 1939–47, while more recently A. P. Thirlwall (1966) has written on the recurrent pattern of regional unemployment. The effect of these variations in the regional demand and supply pattern for labour is to produce higher than average wages in those regions where demand is high and below average wages in those regions with low demand. As a result the industrialist seeking to minimize his labour costs will tend to be attracted to regions where industries are growing only slowly and labour is therefore relatively plentiful.

We have been talking about variations in demand levels at a regional level, but some American workers, most notably V. R. Fuchs, have carried out studies of conditions in the labour market at a town level. Table 2, which is taken from Fuchs (1967) shows that there is a fairly consistent increase in average hourly earnings with city size.

Table 2

Average hourly earnings of non agricultural employed persons in the USA, by city size, 1959. (in dollars per hour)

| | rural | urban places | | standard metropolitan statistical areas | | | |
|---|---|---|---|---|---|---|---|
| | | under 10,000 | 10,000- 99,999 | under 250,000 | 250,000- 499,999 | 500,000- 999,999 | 1,000,000 and more |
| total | 2.00 | 2.12 | 2.23 | 2.39 | 2.43 | 2.56 | 2.84 |
| South | 1.71 | 1.82 | 1.94 | 2.15 | 2.31 | 2.34 | 2.62 |
| Non-South | 2.22 | 2.30 | 2.39 | 2.54 | 2.50 | 2.67 | 2.87 |
| North East | 2.33 | 2.37 | 2.41 | 2.41 | 2.36 | 2.51 | 2.79 |
| North Central | 2.11 | 2.22 | 2.33 | 2.61 | 2.61 | 2.79 | 2.90 |
| West | 2.36 | 2.43 | 2.50 | 2.65 | 2.62 | 2.71 | 2.98 |
| White males | 2.24 | 2.43 | 2.61 | 2.78 | 2.77 | 2.96 | 3.29 |
| White females | 1.45 | 1.49 | 1.57 | 1.65 | 1.69 | 1.82 | 2.00 |
| Nonwhite males | 1.28 | 1.26 | 1.33 | 1.53 | 1.89 | 2.00 | 2.08 |
| Nonwhite females | 0.83 | 0.69 | 0.91 | 0.85 | 1.05 | 1.24 | 1.47 |
| South: whites | 1.80 | 1.98 | 2.14 | 2.34 | 2.46 | 2.54 | 2.86 |
| nonwhites | 1.06 | 0.99 | 0.99 | 1.13 | 1.28 | 1.37 | 1.54 |
| Non-South: whites | 2.22 | 2.31 | 2.40 | 2.56 | 2.52 | 2.71 | 2.96 |
| nonwhites | 1.80[1] | 1.62 | 1.84 | 1.90 | 2.13 | 2.18 | 1.96 |

1   Based on fewer than 50 observations

Source: Fuchs V. R. (1967) 'Hourly earnings differentials by region and size of city' in *Monthly Labor Review*, Vol. 90, pp. 22–6.

Notice that in rural USA in 1959 the average hourly earnings of non-agricultural workers was $2.00 per hour, while in cities of 250,000 to 499,000 people it was $2.43 and in cities of over one million it was $2.84. Part of this variation may be explained by the higher cost of urban living or the higher quality of city labour, but it may also be due to a higher level of demand for labour in cities in relation to the supply and to competition for labour by the large number of firms concentrated in major urban centres. Once more much work needs to be done to ascertain over what size of geographical area similar supply and demand conditions for labour, and hence labour costs, can be said to prevail.

## 2.6   Forces which change the pattern of labour cost variation

We have been considering variations in labour costs as if they presented a static pattern which industrialists needed to ascertain if they wished to take advantage of low labour cost areas. However labour cost variation also has a dynamic element to it and one can consider the factors which lead to CHANGES in cost from one point in time to another. In a perfectly competitive labour market the spatial differences in wage levels and labour costs would disappear. Labour would migrate from low wage areas to high ones so that the relationship between the supply and demand for labour in all areas would become the same and wage levels would be equalized. However, one can argue that in reality there are a whole series of barriers to free movement of labour both occupationally and spatially and there is a whole field of necessary research concerning the nature of these barriers and the strength of the forces attempting to overcome them.

## 2.7   The geographical distance over which labour moves

Most movement of labour takes place on a daily basis over short distances as people travel to and from work. The distance over which they will be prepared to move will naturally vary and will be influenced by the speed and frequency of transport. In the UK it seems likely that most workers would be unwilling to travel more than ten miles or spend more than an hour in travelling, although London commuters are clearly an exception. In areas or countries where motor car ownership is high, travelling distances are likely to be greater because of the greater distance which can be covered in a given time by private as opposed to public transport. In the UK this should produce a series of geographically small labour markets each with their own supply and demand conditions. The fact that labour supply does change very markedly over short distances can be illustrated by reference to a leather firm in Walsall. This firm had, in 1964, three factories in the town all making basically the same products. Two of these factories experienced extreme difficulty in getting labour while the third had no such problems. The factories which were experiencing conditions of labour shortage were located on the south side of the town while the other was located on the north side and on a bus route from Cannock Chase. The Chase area is one which had fairly large numbers of jobs for men in coalmining and engineering but offered few opportunities for female employment. Women were prepared to travel to north or central Walsall in search of work. They were not willing to travel to the south of the town since that involved a change of bus and a consequent increase in travel time. To the south of Walsall lies the great bulk of the Birmingham-Black Country conurbation where demand for labour was high and the number of job opportunities for women large. As a result the factories to the south of the town centre had a more restricted supply of female workers to draw upon.

Not only are market areas for labour small and journey to work distances limited but there is also considerable evidence from studies both in the UK and in the USA that most of the labour force is very unwilling to change place of residence in order to obtain a job. In a study of the mobility of tool and die makers in the USA it was found that over the eleven year period 1940–51 only 8.4 per cent of the 1,712 men interviewed had moved to a new town in search of employment. In contrast 25 per cent had changed jobs. This suggests that while the labour force is mobile to an extent between occupations, the amount of movement from one labour market to another is fairly small. This means that instead of supply and demand conditions for labour, and hence labour costs, becoming equalized from place to place, wide disparities in supply and demand conditions and wage levels are likely to persist for long periods of time.

Some movement of course does take place between labour markets and this differs both in the distance moved and in the degree of permanence of the movement. In general one would expect that as regional differences in wage levels increase, the volume of labour migration will increase and the distance over which migration is contemplated will also increase. One could argue that international migration, for example from India, Pakistan and the West Indies to the UK, would only take place when wage differences were very great. In contrast one would expect migration within the UK to take place in response to rather smaller wage disparities simply because the distance involved in movement is less. Quite a number of attempts have been made to identify causes of interregional migration in Britain of which the work of Oliver (1964) which examines the period 1951–61 is one.

## 2.8 The degree of movement between occupations

While the reluctance of workers to migrate from one labour market to another helps to perpetuate differences in labour costs between areas, differences in costs within areas are maintained by the differences in skills possessed by members of the labour force. The skills required in different industries may be so dissimilar that even within one labour market it is possible to have an acute shortage of labour in some industrial sectors existing side by side with a labour surplus in others. Moreover skills act as a barrier to movement not only within labour markets but between them. Unskilled workers tend to be most willing to move, semi-skilled workers are less willing and many highly skilled workers highly unwilling to move. This is because, with an increase in the spatial specialization of the economy, skills acquired in one labour market in response to a local demand may be completely useless in other markets. For example in the UK there have been considerable difficulties in the redeployment of redundant coalminers because their skills are of little use in most industries where the labour force is expanding. Professional workers such as doctors, dentists, and academics tend to be the exception to the pattern of increasing skill being associated with decreasing geographical mobility. This is because job opportunities are more limited and the effective labour market is virtually national in scale.

## 2.9 Plant relocation as a means of labour cost equalization

So far we have talked of labour migration as one way in which the differences in labour cost and wage levels from place to place can be reduced. Another way for such differences to be reduced is by plant migration. Firms who find that a large proportion of their total production costs are accounted for by labour charges, and who notice that labour costs differ markedly from location to location, may be attracted to low cost areas. It has often been suggested that the lower costs of labour in southern USA has been a major cause of industrial migration of firms from the Manufacturing Belt into states such as Alabama, Georgia and North Carolina.

Once more, however, there are powerful forces working against the relocation of firms from a high cost labour area to a low one. Existing plant and buildings need to be written off and new plant constructed at another location. In addition in the UK workers who are made redundant by a firm's decision to move to a new location have to be paid compensation. As a result the costs incurred in moving to a new location are considerable and may offset any advantages to be gained from lower labour costs. Moreover any firm, when faced with high labour charges, can adopt either a spatial or a non-spatial strategy to overcome these. The spatial strategy involves the sort of plant relocation we have been discussing, while the non-spatial one involves substituting more machinery for labour, so that the firm can operate with a smaller but more highly productive workforce. This process of substituting capital equipment for labour may be an easier and less costly one than that of relocation. As a result firms attempt to overcome high labour charges without relocation whenever they can. However, since labour and capital are not perfect substitutes there is clearly a limit to this sort of operation.

There are other constraints apart from costs on the relocation of firms. Many small firms of under twenty employees are often closely linked to local suppliers of raw materials and serve strictly local markets. Relocation for them would often involve seeking new suppliers and new markets in areas with which the factory owners are unfamiliar. Most small firms depend for their survival on a detailed knowledge of local conditions and would simply lack the expertise to build up new markets and supply patterns. The large firm, in contrast, has greater technical expertise available to enable it to build up new markets and it will often, in the first instance, open up a branch plant in a new location. This enables it to assess the profitability of that plant before deciding on more wholesale relocation. As a result it tends to be only the large firms who are willing to even contemplate relocation from one labour market area to another, although small firms do relocate within market areas.

## 2.10   Other factors affecting labour costs

Migration of workers and migration of firms, however constrained that migration may be, are forces which tend towards the equalization of labour costs between geographical areas. Some factors which produce change in labour cost patterns however, do not necessarily tend towards equalization. For example on a micro scale the industries in a particular area will change in their degree of prosperity and will change their demands for labour as the demand for their products changes. In addition technology may reduce the demand for labour by making it economic to substitute capital, in the form of machinery, for workers. This will affect not only the overall demand for labour but also the availability of overtime and part time working.

A long-term change in the supply and demand relationship for labour in a particular area can of course be produced by population. This is the major cause of changes in the size of the labour force. For example over the period 1963–73 the working force of the USA is expected to grow at over four times that of the rate of the UK and the six original EEC countries. This means that the USA could expect to experience a far greater rate of industrial expansion than will be possible in Europe. This is because Europe will quickly run up against manpower shortages and undergo an inflationary spiral of wages and labour costs. This sort of long-term change in the supply and demand relationship for labour takes place at all geographical scales from the town to the continent.

Another factor which may cause labour costs to change through time is the action of governments. The general impact of government action on location patterns will be discussed in Unit 5, so two brief examples will suffice here.

In the UK the Labour Government's granting of a regional employment premium to firms in the development areas effectively meant that firms in manufacturing industry received a cash grant for every worker on their payroll. This quite naturally lowered their labour costs. The other example, already encountered, is that of the impact of state labour legislation in the USA on the cost of labour and the flexibility with which that labour can be used.

## 2.11    The impact of labour costs on location – a summary

The reasons why labour costs vary from place to place and the way in which those variations can change through time have been discussed at length for two reasons. Firstly, for many firms labour costs are now one of the largest and most variable costs with which they are confronted. Secondly, a tremendous amount of work has been done on various aspects of labour cost variation, and this is difficult to review briefly.

Perhaps the discussion of labour costs can best be summed up by reference to Figure 1. This map is one of a series produced by D. M. Smith and shows how wage levels differ from one part of the USA to another.

Figure 1

Average hourly earnings of production workers in manufacturing industry in the USA, December 1966.

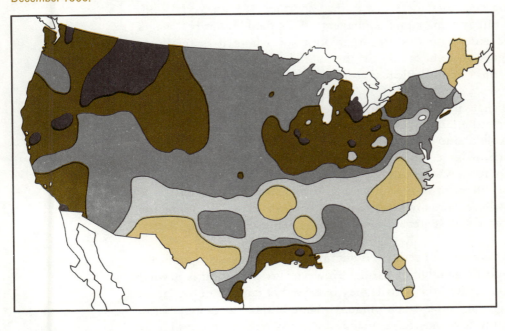

average hourly earnings ($) 1966

| | | | | |
|---|---|---|---|---|
| 1·85–2·23 | | 2·61–3·00 | | 3·38–3·76 |
| 2·23–2·61 | | 3·00–3·38 | | |

Source: Smith, D. M. (1971) *Industrial Location. An economic geographical analysis,* New York & Toronto, John Wiley & Sons.

The geographical study of labour costs is concerned with the production and explanation of such maps and with the way in which mappable patterns change through time. It is then concerned to examine the way in which such mapped variations can explain the variations in the production costs of firms from location to location and the way in which such cost variations affect industrial location decisions.

## 3  Enterprise

Enterprise as a factor influencing the geographical variation of production costs is generally regarded as being relatively unimportant. The term can be defined as meaning the managerial skill needed at a policy making level in the running of a modern firm. Clearly top management will be concerned with location decisions, with the mix of the factors of production, with the scale of operations and with marketing policy. The industrial manager has the task of detecting and responding to spatial variations in the cost of land, labour or raw materials and it is the decisions of large numbers of managers and business-men which determine the industrial location pattern. Naturally business decisions must play a central role in any modern theory of industrial location, and these are discussed at length in Units 6 and 7. What we are concerned with here is the very much more restricted problem of whether the cost of obtaining managers or decision takers differs significantly from place to place.

Tackling this problem is difficult because the amount of work actually carried out into the cost and availability of enterprise at different locations is very small indeed. Generally it is assumed that managers are equally easy to obtain in virtually all locations and that therefore wage levels for managers do not show any systematic geographical variation. Chinitz (1961), however, argues that those economists who study international variations in economic growth rates often stress that expertise is unequally distributed between countries and that in many developing countries there is a critical shortage of managerial skill. Presumably, other things being equal, managers will be paid higher real wages in countries where enterprise is scarce than in those where it is relatively plentiful. Chinitz goes on to argue that if the availability of enterprise differs between countries there is no logical reason why it should not differ within countries.

Chinitz suggests that in those industries where there are large numbers of small uni-plant firms the families of managers are more likely to be involved in and knowledgeable about industrial decision taken than are the families of executives who work for large industrial corporations. On Chinitz's hypothesis it is the small family firms who not only produce managers for their own needs but who also supply them to large companies. It follows from this that those areas of a country where there are large numbers of small firms will also be the areas where managerial skill (enterprise) will be most easily available and hence cheapest.

Chinitz's hypothesis has not been tested empirically although he argues that in principle it should be capable of testing. However great caution is needed in examining an argument of this sort. It may be true that families of managers of small firms do get to know a fair amount about business decision taking but much of that knowledge may be quite inappropriate to the sort of problems which face the managers of large firms. In addition the fact that there are more opportunities for people to become managers in areas where small firms predominate, does not mean that managerial ability is concentrated there. It may be that the bulk of people recruited as managers now have no managerial background at all but have gained their knowledge by means of university and vocational training courses.

An alternative approach to the assessment of spatial variations in the supply of managers is to study their locational preferences. D. M. Smith suggests that in Britain it is easier to obtain high quality managers in London and the South-East because many of them value the social amenities which London provides and have an image of the north of England as a region which is physically unattractive, climatically unpleasant and lacking completely in social amenities. If managers do indeed have this image then it may well be necessary for firms located in the north of England to offer higher salaries in order to persuade them to venture there. In contrast, however, firms in London may have to pay high salaries to their executives simply to offset the higher cost of living in London. If this is so then the wage costs of firms in London may be exactly the same as those in any other part of the country.

Again detailed empirical evidence seems to be lacking although Smith does suggest that managerial perception in the USA encouraged the migration of enterprise to Florida and California and aided the process of industrial growth there. Moreover if Galbraith (1967) is right in arguing that industry in the developed countries is being increasingly run not by factory owners but by a professional executive group, then the perception of different areas by these executives and their willingness to move from area to area could have an important effect in producing geographical differences in the cost and availability of enterprise.

One final suggestion made by Smith on possible geographical variations in the availability of enterprise requires mention. He argues that if firms wish to recruit a group of managers, each one with specialist skills, they are more likely to be able to do this in large urban areas than elsewhere. This is because in areas with large concentrations of population there are bound to be people with specialist managerial skills. If this is so then firms will tend to locate either in or near large cities and (leaving the cost of living out of account for a moment) should be able to pay them lower salaries because there is no need to induce them to change their place of residence.

This suggestion, like all the others, seems to lack the backing of empirical work. One can only reiterate that geographical variations in the cost of enterprise are generally assumed to be unimportant in affecting the total operating costs of factories in different locations, and that a considerable amount of empirical work needs to be done. However, future work or enterprise is likely to be made increasingly difficult by the fact that head offices of manufacturing firms are becoming separated from the factories they manage. Once more it is necessary to stress that the discussion of enterprise does not suggest that all executives are paid similar salaries, or that their degree of competence does not vary. What it does suggest is that such variations as do occur do not seem to produce an easily recognizable geographical pattern.

## 4   Capital cost variations

Capital is yet another factor the cost of which can vary from place to place and hence contribute to geographical variations in the production costs of manufacturing firms. The capital involved in industrial production consists of two types, fixed capital and financial capital. Fixed capital represents investment in buildings and equipment which is thus relatively immobile geographically. Financial capital, in contrast, represents investment funds in the form of dollars or pounds, and this is relatively mobile between locations. The degree of mobility depends upon the geographical scale (i.e. national or international), and the degree of development of the country concerned. Within advanced countries capital funds are more mobile between areas than within underdeveloped areas.

### 4.1   Financial capital

Financial capital for industry is usually regarded as being available at similar interest rates at all geographical locations. It is thus regarded as being unimportant in causing costs of production to vary from place to place and as having an insignificant influence on location decisions. Careful thought will quickly show that the assumption about similar interest rates is incorrect. On an international scale interest rates on capital borrowed for industrial development

clearly do vary from country to country according to the demand for investment funds in relation to their supply and to the amount of control on the supply
of funds being exerted by national governments. In addition interest rates
within countries will vary. A firm trying to raise capital for investment outside
the national area, will expect to pay higher interest charges on borrowing
for industrial schemes in countries which are politically unstable or where the
probability of expropriation of foreign assets is high than on borrowings for
investment in Western Europe. Interest rates for investment schemes within a
national area may also vary. Banks might for example demand higher interest
rates for loans to build factories in a depressed region of the country or in
locations away from major consuming centres for the factories' products. In
such cases the decision of whether or not to charge higher interest rates would
depend upon the type of industry and the bank's assessment of the probability
of the proposed schemes being successful. Both Chinitz (1961) and Beckmann
(1968) seem convinced that interest rates, and hence the cost of obtaining
financial capital, do differ from one geographical location to another. They
argue that these variations are produced by geographical differences in the
risks involved in industrial investments in different locations. Again empirical
evidence on interest rates and risk levels is largely lacking. However as an
example of risk variation one might think of the higher risk of investing in
factory buildings in Northern Ireland as opposed to elsewhere in the UK.
While civil strife persists the possibility of arson must be rated higher than elsewhere and this may be reflected in interest rates on borrowings for investment
there.

Variations in the cost of financial capital will affect firms of different sizes
in different ways. Generally, the national or international company will
not have to bear higher interest charges on capital invested in high risk locations. This is because such firms are usually able to finance industrial development either out of retained earnings or by resort to the stock exchange. As a
result it will rarely need to seek investment funds specifically for high risk
projects. Even where this is necessary the risk involved in lending money to
large firms is usually fairly small. In contrast firms with under 100 employees
usually lack sufficient retained earnings and access to the stock exchange as
a means of financing new investment. As a result they are much more heavily
dependent on bank loans than are large firms. Chinitz argues that small
firms find it very difficult to obtain finance outside their own local area. This,
he says, is because only local banks will know sufficient about a firm seeking
a loan to know the degree of risk involved. Banks even as little as twenty miles
away would lack such information, be unable to assess the risks in making a
loan and therefore either refuse to provide any cash or demand prohibitively
high interest rates.

Banks, of course, are not the only source of finance for small firms. In the
USA there are a large number of state and local authority backed loan organizations who provide capital at low interest rates for small firms. For example
in 1963 seven American states had schemes which guaranteed the repayment
of first mortgage industrial loans made by private lenders. Of the loans made
under these schemes 93 per cent of the number of loans and 60 per cent of the
dollar volume went to plants with under 500 employees, but 50 per cent of the
dollar volume went to plants with under 200 employees. The geographical
distribution of schemes of this sort is irregular and has the effect of making
interest rates on loans much cheaper in those areas where such schemes exist,
than in areas where they are absent. More detailed discussion of the role of
governments in modifying industrial costs will be found in Unit 5.

Despite the arguments of Chinitz and Beckmann that the costs of financial
capital do vary from place to place, and despite the uneven distribution of loan

guarantee organizations, empirical evidence is largely lacking. However Smith (1971) does cite some empirical work carried out by the Fantus Company. The company calculated the average annual financing cost, including principal and interest, for industrial operations in a series of American and Canadian cities. They found that interest rates varied in 1960 in thirty four cities between a high of 9.31 per cent in Charlottesville, Virginia to a low of 7.92 per cent in Cynthiana and Elizabethtown, Kentucky. There seemed to be no discernable relationship between interest rates and either geographical location or city size. We are thus forced to the conclusion that while intuitively there are good reasons for believing that the cost of mobile capital does vary from location to location, what little empirical evidence there is suggests that variations do not produce any clearly, discernable geographical pattern. There seems to be no factual evidence as yet to allow us to refute the commonly held belief that the costs of financial capital have little influence on industrial location decisions or production costs, apart from at the international level.

## 4.2   Fixed capital

Fixed capital is usually defined as meaning factory buildings and machinery. It is termed fixed because it is immobile geographically and while a machine may be moved from one place to another one could scarcely imagine the same happening with a factory building. Costs of machinery naturally differ from supplier to supplier but because most manufacturing firms buy machinery relatively infrequently it can be assumed that a firm will not feel compelled to locate near a producer of cheap equipment. It must be confessed however that at present there has been no attempt in Britain to map the variations in the cost of machinery and to see if they produce any comprehensible geographical pattern.

Geographical variations in the costs of industrial buildings have received rather more attention but not much more. Such work that has been done can be divided into that which is concerned with the cost of building new factories and that which examines variations in the prices and rents of existing premises. Smith (1971) has used the Dow Building Cost Calculator for 1969 to map the variations in industrial building costs in the USA. These results are shown in Figure 2 overleaf.

The Building Cost Calculator gives a basic cost per cubic foot for a series of different types of factory buildings. This cost includes construction costs but excludes the cost of putting in foundations, design fees and builders' profits, all of which could well vary from one geographical location to another. The map shows that the lowest building costs are to be found in the Old South in states such as Georgia, Alabama and North and South Carolina and are highest in the north-eastern part of the USA, where most of the nation's manufacturing industry is concentrated. Smith points out that the general pattern of variations in building costs seem to follow that of variations in wage rates (Figure 1). He tentatively suggests that variations in the wage rates of building workers from one part of the USA to another may be the major cause of building cost differences. Considerable empirical work still needs to be done both in the UK and in the USA, on how and why industrial building costs vary from place to place and what effect such variations have on the location decisions of manufacturing firms. It may well be that in particular areas building costs are similar not only because of similarities in the levels of building wages but also because builders, via their trade associations, have decided what sort of price it is reasonable to ask for industrial work.

The cost of buying or renting existing industrial premises seems to vary with their age and with the supply of such buildings in relation to demand. By and large the purchase price or annual rent of older factory buildings will

Figure 2

An industrial building-cost surface, derived from relative cost indexes for 181 metropolitan areas in the USA.

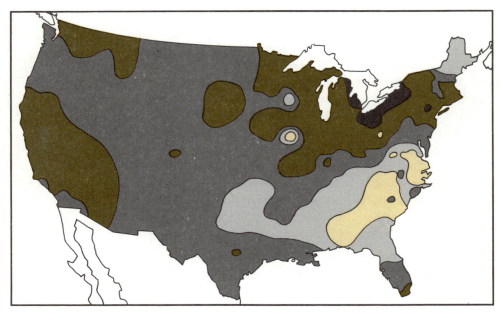

index 1969 (base cost = 1·00)

| | | |
|---|---|---|
| 1·85–2·01 | 1·54–1·70 | 1·23–1·39 |
| 1·70–1·85 | 1·39–1·54 | |

Source: Smith, D. M. (1971) *Industrial Location. An economic geographical analysis,* New York & Toronto, John Wiley & Sons.

be much less than for new ones. In Birmingham for example it emerged that the annual rent for one hundred year old industrial buildings in central Birmingham was only £100. In contrast the rents for new industrial premises of similar size and in a similar location cost between £250 and £3,000 a year. In most cities both in the UK and the USA the oldest industrial buildings tend to be found near the urban core, while an increasing amount of new building is being carried out on the urban margins. As a result the average price and rent of industrial premises tends to increase from the centre of the town outwards.

The supply of industrial buildings in relation to the demand for them is a function of the relationship between the birth rate of plants or factories, their death rate and the rate of new industrial building. Birth rate is a term for the number of new firms or branch plants of existing firms wishing to open premises in a particular area. It determines the level of demand for industrial buildings, and tends to increase when the trade cycle turns upwards. The supply of existing industrial buildings for sale or rent partly depends on the number of firms or branch plants which are going out of business and are thus vacating premises. This is referred to as the death rate of firms and tends to increase as the conditions of trade deteriorate. The supply of premises also depends on the number of firms who are having new factories built and are planning to vacate their existing premises when these are completed. The rate of new building is usually highest when terms of trade are good. In general terms as the birth rate of firms and hence demand is increasing, the death rate is falling so that a growing proportion of demand has to be met by firms leaving premises to move to newly built ones. In those circumstances the total supply of industrial buildings in an area will increase, but probably at a much slower rate than demand. This will cause the purchase prices and rents of existing factory premises to increase.

Interpreting this process geographically shows that in those areas which have a higher than average birth rate of firms and a lower than average death rate the cost of industrial buildings will tend to be higher than elsewhere. This general relationship is however modified by regional differences in average firm size (measured in terms of square feet of factory and storage space) since an area with an apparently below average birth rate may in fact be one where the amount of floorspace per firm is higher than elsewhere. In general, however, one would expect high birth rates, low death rates and high building rents in areas which are dominated by industries which are growing faster than the national average.

Empirical evidence concerning the variation in rents and prices of existing industrial buildings is limited, although Smith (1971) does provide a review of the situation in Lancashire. He argues that in the old cotton producing areas the rapid decline of the textile trades has resulted in many mill premises falling vacant. These can be obtained at very low prices. Work done by Holt (1964) showed that of 622 mills closed in the region during the period 1951–62, 414 had been converted to other uses, 63 had been partly converted and only 145 remained unused. Smith points out that similar colonizations of old mills and warehouses have taken place in New England, and in the lacemaking districts of Nottingham. Thus in terms of our earlier discussion death rates in the textile industry have been high because of the decline in the demand for the industry's products. This has produced a situation where the supply of industrial premises has exceeded the demand, prices have been held down and new firms have been attracted to these areas by the low cost of factory premises.

In conclusion it can be argued that both the costs of new and existing industrial premises do vary from place to place and there is some evidence to suggest that this influences the detailed location decisions of firms.

## 5  Energy

The costs of energy required to carry out industrial processes are generally assumed to play a relatively unimportant part in the location decisions of industrial firms and to make little difference to their production costs in different locations. This is because fuel costs are likely to remain virtually constant over quite wide geographical areas. The development of a national electricity grid in Britain for example meant that power from any given station could be economically transmitted to most parts of the country. As a result the varying production costs of power stations ceased to have an impact on the price of electricity in the areas surrounding those stations. The creation of the national grid meant that consumption might take place many miles away. In Britain the responsibility for selling electricity rests with a series of regional boards and they are permitted, within limits, to vary the price of industrial electricity according to their differing distribution costs.

In the USA a study by J. B. Lansing in 1945 showed that fuel costs for 197 steam power stations with 5,000 kilowatts capacity or more showed quite distinct regional patterns. Costs, measured in cents per million Btu, were highest in New England and Florida and lowest in a block of states running from Missouri to Texas. This low cost group of states, where fuel costs averaged 6–12 cents Btu, were either in or adjacent to areas of oil and natural gas production. Only slightly higher fuel costs were found in California where oil is produced, and in the coalmining states of Kentucky, West Virginia, Illinois, Wyoming,

Colorado, Pennsylvania and Alabama. Assuming that electricity prices followed closely the pattern of production costs, some indication of the varying costs of power to the industrial consumer can be obtained.

In general terms regional variations in electricity prices are gradually becoming less important as new links in the electricity supply network are added and permit the transfer of power over even greater distances. This process of price equalization is being aided by economies in transmission costs. A similar pattern is emerging for gas supply. Until about ten years ago in Britain most gas was produced locally and consumed within twenty and often even ten miles of its production point. Since then plants using the ICI reformer process, which obtains gas from oil, has begun to supply gas on a regional basis, while natural gas is increasingly being pumped to many areas by means of a pipeline grid. This has had the effect of evening out variations in gas prices produced by local differences in production costs.

While energy costs are relatively unimportant for most industries today, there are a few industries which demand vast quantities of energy at low price. In such industries factories are often found side by side with plants producing cheap power. For example, an aluminium smelter has been suggested for the island of Anglesey because it could obtain power from the new atomic power station at Wylfa. An aluminium smelter has already been built at Fort William in Scotland, and takes advantage of the very cheap supplies of hydro-electric power produced there. Similar supplies of cheap power have led to the growth of an aluminium industry in the Pacific North West of the USA, in the Tennessee Valley and in the French Alps.

Isard and Whitney (1950) point out that at the time they were writing something between 18,000 and 22,000 kilowatts of electric power were needed to convert alumina (refined ore) into one ton of aluminium ingots. The work of Krutilla (1955), however, provides a warning against assuming that large power users will always locate at the source of cheapest power. He compared the cost of producing aluminium in the three regions of the Tennessee Valley, the Texas Gulf Coast and the Pacific North West. He found that while the cost of 18,000 kilowatts of electricity was only 49·50 to 58·60 dollars per ton in the north-west USA as compared with 72·00 to 78·90 dollars in Texas, the cost of all other factors of production were lower in Texas than elsewhere. Although he concluded that production and marketing costs for aluminium were virtually identical in Texas and in the Pacific North West of the USA, he found that the bulk of new capacity had been located in the higher cost fuel region of Texas. In contrast, more recent developments in the aluminium industry again show energy orientation, with major plants being built near to coal-fired power stations along the Ohio river.

The locational hold which power once exerted over the whole of industry was very considerable when coal was the main source of energy. At the time of major industrial growth in Britain and the USA coal was expensive and difficult to transport. As a result major concentrations of industry began to be formed on or near coalfields. Considerable investment was made both in factory buildings and houses for the labour force. Specialized industrial complexes emerged where the finished product of one industry became the raw material for another. Once such large concentrations of fixed capital had been created they acted as a powerful force against relocation. An industrialist wishing to take advantage of the gradual slackening of the locational grasp of fuel had to bear the cost of building a new factory, of moving his workers and of loosing the close ties with suppliers and consumers which he had in his concentration location. The consequence of these pressures against locational change has been that major concentrations of industry still survive, and yet owe their original rationale to access to cheap energy from coal. Thus we can see that if energy plays a relatively minor part in industrial location decisions today, it has nevertheless left us an industrial location pattern whose major features were determined by the availability of fuel.

## 6   Raw materials

The impact of raw materials on industrial location has long been examined both by geographers and space economists. The theorist Alfred Weber, whose work was discussed in the Foundation Course, *Understanding Society*, and is also described in Unit 3 of the block, spent a considerable time examining how the costs of assembling raw materials at a factory could be minimized. His work highlights the fact that when any raw material arrives at a factory its cost is a function of two things:

(a)  the cost of producing or mining it;
(b)  the cost of transporting it to the factory.

Section 7 will deal at length with the impact of transport on production costs and factory location, so the discussion which follows is confined to a consideration of how the production or mining costs of raw materials differ from one location to another.

At this point the concept of raw materials needs to be examined more closely. Raw materials are usually defined by economic geographers as the material upon which a factory carries out a productive operation. These materials have to be brought from their producers to the factory and are therefore often referred to as production inputs, whereas the finished product of which they form a part is referred to as an output. It will be clear that given this definition all factories will have production inputs but that most of these will not be 'raw materials' in the conventional sense at all. Some inputs will naturally consist of unprocessed material such as iron ore or logs, but the input of most firms will consist of semi-finished products, or even finished manufactured products which are simply assembled at the factory. Car assembly plants are a good example of factories which have finished products as inputs.

The factors which influence the cost of true raw materials differ somewhat from those which affect semi-finished and finished products so they will be considered separately.

### 6.1   Raw material costs

Most raw material inputs for manufacturing industry consist of minerals although some industrial sectors such as food processing, the traditional sectors of the textile industry and timber processing depend upon animal and vegetable products. The cost of mineral extraction is a function of geological conditions, labour costs, degree of mechanization or technical skill and scale of operations. These factors interact with demand levels to determine the ex-mine price. Demand levels also help to determine the location and scale of mining operations. Geological conditions influence the cost of mineral extraction because they can determine the speed and method of mining. For example if a particular mineral resource dips away almost vertically beneath the surface of the earth the costs of extraction are usually much higher than for a similar resource which is horizontally bedded near to the earth's surface. Steeply dipping strata requires the sinking and servicing of deep mine shafts, ventilation becomes a problem and the cost of winding raw materials to the surface increases. In addition the thickness of a particular deposit is important because a thick bed or vein usually permits the use of mechanical cutting equipment, which reduces the cost and accelerates the rate of mining. Chemical composition of resources too is important. Where the material being mined has a high percentage of the required mineral and a low percentage of waste material, the cost of mining one ton of the mineral will be less because less waste will have been mined. The iron industry provides a good example of variations in chemical composition of raw materials. In Britain alone the iron content of ores varies from 55 per cent in Workington and Barrow-in-Furness down to 20 per cent in the marlstones of Leicestershire and Lincolnshire.

Variations in labour cost from place to place also play a part in determining the costs of mining. Wage levels, and hence labour costs, in ore exporting countries such as Angola and Mauritania are substantially lower than those in Minnesota or Northamptonshire. Moreover our earlier examination of variations in wage levels suggests that even within countries wage levels and hence wage costs are likely to differ from place to place.

Technology too, and particularly changing technology, can have an impact on production costs at particular locations. An illustration from recent developments in iron mining and manufacture should help to illustrate this. The Mesabi area of Minnesota was an important iron mining area, but by the 1930s after a century of mining, the high grade ores were becoming exhausted and only low grade ores remained. At the end of the 1930s however a process known as pelletization was evolved which permitted the upgrading of the iron content of ore at the mining site. Over the last ten years this process has become a commercial proposition, ores are being upgraded and are in many senses better than naturally rich ores. Rocks with an iron content of under 40 per cent can be converted into ore pellets which contain 64–68 per cent iron. The introduction of pelletization to Mesabi has permitted the working of ores which were previously completely uneconomic. Already new technical developments suggest that ore pellets could be produced with 95 per cent iron content. As pelletization spreads whole new material sources are being opened up as low grade ores become worth exploiting. Moreover from the point of view of iron consuming firms the geographical distribution of ore suppliers and the prices they charge are being modified. Similar technical changes are occurring in other industries and are altering the pattern of raw material supplies.

Pelletization also throws some light on the fourth factor which affects the ex-mine price of minerals, namely scale of operations. By and large the bigger the scale on which a mining firm can operate the more mechanization is possible and this has the effect of reducing the cost per ton of the mineral being extracted. However the corollary of this is that the amount of initial capital required to commence mining operations increases. For example prior to the introduction of pelletization the initial investment needed to start an iron mine was around $10 per ton of annual capacity, while now it is $30 per ton of annual capacity in the USA and Western Europe and $50 per ton in remote areas of Canada and Australia. This sharp increase in basic capital costs has the effect of excluding small producers from the market and of concentrating production in the hands of a relatively few large producers who market their products over a wide geographical area.

Having examined some of the factors which cause the cost of the production of minerals to vary from one geographic location to another it is necessary to consider the impact of the level of demand for minerals on cost and location patterns. To the extent that an increase in the level of demand for a particular mineral enables mines to operate on a larger scale with more mechanization it has the effect of reducing production costs. Since prices to the consumer usually rise at the same time this provides little comfort for the manufacturing firm. However another effect which demand can have on mineral exploitation can be of more use to the firm which purchases raw materials. As the demand for a particular mineral rises, price also rises. This has the effect of making some locations of raw materials economically viable where previously their costs were too high. If mining operations then begin, the geographical pattern of mineral exploitation will be modified and some consumers may find raw materials being worked very much closer to them than previously. This offers the potential of savings in transport costs.

In summary, the consumer of raw materials may find himself able to choose between a series of suppliers, all of them with different production costs and prices. In assessing which supplier to patronize he has to consider not only the ex-mine price of the raw material but also the transport cost of the material from the mine to his factory. This effectively limits the geographical range within which he can search for suppliers. It naturally follows that in some locations the cost of raw materials will be less than in others, and that a firm wishing to minimize its material costs would be attracted to these locations. This is easiest for a firm to do if it has only one major raw material input, but if it has several it needs to seek a location where the total assembly costs of all raw materials is reduced to a minimum.

## 6.2    The cost of finished or semi-finished products

The factors which affect the price of manufactured 'raw materials' in different locations include many factors already discussed in this part of the block. The firms who supply semi-finished or finished goods have their own production costs determined by labour charges, the cost of power, capital, enterprise, land, transport and the cost of their own raw material inputs. The price they charge for their goods will depend upon these factors, on the level of demand and on the type of pricing stategy they adopt. Some firms charge consumers for the ex-factory price of the goods plus the cost of transport to the consumer. In such circumstances there is an incentive for consumers to choose suppliers who are located near them, or to choose deliberately a location for their factory close to a major supplier. Smith (1971) however argues that this sort of pricing is becoming increasingly rare and that most producers of 'raw materials' simply quote a uniform delivered price which on average gives an adequate profit level and covers transport costs.

Examination of 'raw material' costs reveal that in studying the geographical variations in production costs of firms, one is really examining a chain of costs. The costs of producing raw materials help to determine the price which a raw material consumer will have to pay for his production inputs. The raw material in turn will become one factor in determining the production costs of that consumer and will help to determine the price he asks in turn for his product. Thus the geographical variation in the cost of semi-finished or finished 'raw materials' will be directly dependent upon the production costs and pricing policy of the producers of those goods.

## 7    Transport costs

Economic geographers have for long been interested in transport costs and their influence on industrial location. Reference to Unit 3, Location Theory, will show that Weber developed his entire theory of industrial location from the idea of the cost of carrying 'raw materials' to the factory and of finished goods to the market. The price which any manufacturer has to pay either for his 'raw materials' or to ship his finished products to the market will be a function of two transport elements:

(a) Line haul charges.

(b) Terminal charges.

Line haul charges are those made for actually transporting goods, while terminal charges are the costs of handling cargo at its origin and destination.

## 7.1  Line haul charges and terminal charges

Line haul charges are often regarded as increasing with the distance travelled so that a manufacturer wishing to have goods carried 10,000 miles would expect to pay more than one who wanted transport over only 500 miles. This is simply because actual transport costs in terms of total fuel, wages and other costs will be greater. This does not mean however that line haul charges alone would cause a change in the cost per mile to occur. However when terminal charges are added in, a rather different pattern emerges. The costs of handling goods at their origin and destination remain constant irrespective of distance travelled. These fixed charges are of course spread over much greater distances on long journeys than on short ones so that actual cost per mile progressively falls as distance increases. This proposition can be illustrated by a 'manufactured' example. Imagine that the cost of actually carrying goods was 5p per mile. Then in our example of goods carried over 1,000 miles and 500 miles the total cost of carriage would be £50 and £25 respectively. Imagine that the terminal charges on each consignment amount to £25. The total cost of shipping each consignment will then become £75 for carriage over 1,000 miles and £50 for carriage over 500 miles. Notice that the freight rate per ton mile will become 7·5p for goods carried over the longer distance but 10p for those carried over the shorter distance. Expressed algebraically the freight rate per ton mile would be as follows:

$$r = \frac{a + bx}{x}$$

where $r$ = freight rate per ton mile on any given consignment.
$a$ = the fixed terminal charges.
$x$ = the total distance travelled.
$b$ = line haul charge per mile.

Given this situation we would expect the total cost of carrying a given consignment of goods over different distances to vary in the way shown in Figure 3. It will be clear from the graph that line haul charges per mile remain constant as distance increases and that terminal charges form a declining proportion of

**Figure 3**
Increase in total transport costs with distance travelled.

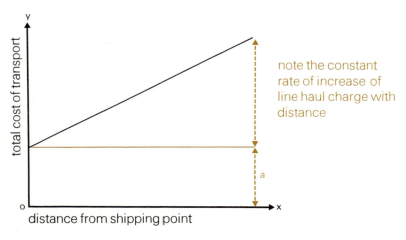

note the constant rate of increase of line haul charge with distance

**Source Figures 3 and 4: G. Edge.**

total costs. When very great distances are travelled the proportion of costs accounted for by terminal charges becomes so small that freight rates become virtually a reflection of line haul charges.

$r = f(x)$. This means that $r$ (freight rate) is some function of $x$ (the total distance travelled).

Applied to our simple equation we get:

$$\lim_{x \longrightarrow \infty} f(x) = b$$

This means that $f(x)$ can be made as numerically close to $b$ as you like by making $x$ (distance travelled) large enough. In detail we take the equation:

$$f(x) = \frac{a + bx}{x} \quad \text{and simplify it}$$

This gives $f(x) = \dfrac{a}{x} + b$

Assume that you want $f(x)$ to be within $\dfrac{1}{1000}$ of $b$ then let $x = 1000a$. Putting this value of distance into the equation we get:

$$f(x) = \frac{a}{1000a} + b$$

which equals $f(x) = \dfrac{1}{1000} + b.$

Students who cannot follow the mathematics should not be too disturbed as the preceding paragraph makes the same point in words.

So far we have assumed that line haul charges are linear and that they will increase continuously and at a constant rate as distance travelled increases. This situation is shown in figure 3. The freight rate structure shown in figure 4(a) is a more general form of freight structure, where freight rates are assumed to rise continuously with distance but at a gradually decreasing rate. Under this system the cost of carrying goods long distances is less than one in which line haul charges per mile are constant. However both the freight structures shown in figure 3 and figure 4a are called 'mileage rates' because under both systems freight charges rise continuously with distance travelled.

It is often assumed that transport costs will increase continuously and at the same rate as goods are carried further and further from the producing plant in any direction. Alexander, Brown and Dahlberg (1958) describe this particular assumption as a complete but widespread fallacy. They found in a study of

**Figure 4**
Mileage and blanket freight rates. (a) Mileage rates. (b) Blanket rates.

(a) mileage rates                    (b) blanket rates

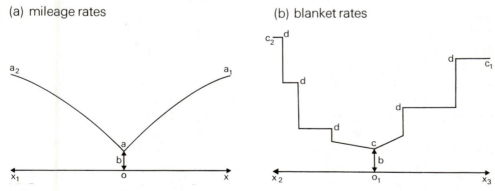

$o$ = loading point of goods.
$b$ = terminal charges.
$ox, ox_1$ = distances from the loading point $o$ to the destination of the goods.
$aa_1, aa_2$ = level of transport charges.
$cc_1, cc_2$ = level of transport charges from loading point $o_1$.
$o_1 x_2, o_1 x_3$ = distances from loading point $o_1$ to the destination of the goods.
$d$ = steps in the freight structure.

freight rates in Wisconsin that although freight charges did increase with distance travelled they did not increase at a continuous rate and that they differed markedly according to the direction in which goods were being transported. They found that along most transport routes, and particularly along

The pattern of freight rates found in Figure 4(b) are usually referred to as blanket rates because over large areas the rate remains the same. Stepped charges are often found in reality because they combine the advantage of being able to charge for goods according to the distance over which they are carried with an avoidance of the problem of having to quote a freight charge for every possible journey between any two places, which is the major drawback of the continuous rate system.

In addition to finding that freight rates in Wisconsin were stepped, Alexander, Brown and Dahlberg also found that steps differed in height, frequency and distance from the loading point according to which direction goods were being carried away from their origin. They found too that in some areas, where different methods of transport were in competition, freight charges were deliberately held down.

So far we have encountered two types of freight rate. Firstly there was the mileage rate, where charges were directly related to the distance over which goods were being carried. Secondly we have encountered the blanket rate. This second type of rate not only encourages firms to choose their suppliers and locate their factories in such a way as to reduce transport costs but it also encourages them to export and import goods across as few steps as possible. There is, however, a third type of freight rate which has no effect on location decisions at all. This is what Smith (1971) calls the 'postage stamp' rate where a uniform transport charge is made irrespective of the distance covered. This sort of rate is of course easy to operate and removes incentives for firms to locate near to their suppliers or markets.

## 7.2  Other factors affecting transport charges and choice of transport

Apart from the structure of freight rates there are a number of other factors which affect the cost of transport and the type of transport used. Different freight rates are quoted on different types of goods. By and large fragile goods and those which require a great deal of careful handling, such as precious and dangerous goods, will be more expensive than goods which are easy to handle and cannot suffer damage. Chinitz (1960) argues that since the 1920s in America the cost of transporting raw materials has fallen relative to other costs and Barloon (1965) stresses that the output of US manufacturing industry is increasingly in the form of sophisticated domestic and industrial equipment. Since this equipment is more easily damaged and more expensive to transport than raw materials it may explain why a growing proportion of firms in the USA are locating near the market.

The size of shipment of goods also influences costs. By and large more is charged per ton on goods which fail to fill a complete barge, railway truck or lorry, since the transport of goods in such quantities involves underutilization of the carrying capacity of the transport system. The gradual replacement of the barge by the railway and of the railway by the lorry has produced a situation where the penalty for shipping goods in small consignments has diminished. Moreover each successive innovation in transport has produced a situation where terminal costs have fallen both absolutely and relative to other costs. Also the flexibility of the transport modes has increased. Lorries for example have lower terminal costs and can go to more places than can railways, which in turn have the edge both in terms of terminal cost and flexibility over barges. One might wonder in those circumstances why rail and barge transport (which is still important in the USA) are used at all. However the line haul charges for rail travel are lower than for lorries, and lower still for barges. When line haul and terminal charges are added together a pattern emerges which shows that lorries are most economic over short distances, railways over intermediate distances and barges over long distances. This means that for firms located near to the market for their products, lorries which can deliver small consignments quickly and easily are the most economic transport.

## 7.3   Effects of transport on industrial location

There seems to be little dispute that the transport costs which particular firms will have to bear will differ from place to place. There is however controversy over the amount of influence these cost variations have on industrial location decisions. Certainly going back to the Industrial Revolution in Britain the use of coal as the major fuel for industry, combined with the difficulties of transporting it, had a major influence in fixing industrial development on the coalfields. A considerable amount of industrial building was carried out and the industrial complexes survive to this day. During the period since the 1950s transport costs have fallen and the methods of transporting goods have increased. Coal has ceased to be a major source of industrial fuel and has been replaced by electricity, which is available almost anywhere at virtually the same price. Despite this many firms remain in major industrial areas and firmly resist any attempts to relocate them on the grounds that they have close production links with firms in the same area. Would these firms, if they were relocated, have to pay unacceptably high transport costs and accept unsatisfactory delivery times from their suppliers, or is their opposition based simply on a reluctance to move to a new area irrespective of the advantages or disadvantages? Barloon (1965) argues that transport costs no longer play a significant part in determining the location of factories. Chinitz (1960) in contrast sees transport costs as playing a continuing part in influencing industrial location.

## 8   Land costs

Discussion of the variation in the operating costs of firms from one location to another has so far centred on how the production and transport costs are likely to be affected by the choice of location. However, in choosing a location for a factory, land has to be purchased, and this too differs in price from one part of a country to another and from one area of a city to the next. Land prices vary on two quite distinct geographical scales. Firstly they vary at a regional or national level and secondly they vary at the detailed local level.

## 8.1   National and regional variations in land costs

At a national level quite marked variations in land cost can be detected. D. M. Smith (1971) gives figures for the regional variations in cost of residential building land in England and Wales in 1966–7. They show that on average £27,783 per acre was paid for land for houses in London as compared to £15,328 in Essex and Hertfordshire, £6,135 in the Midlands and a mere £2,296 in Wales. This suggests a national pattern of land values which are at a peak in London and then gradually fall away to the west and north of the metropolis. This general pattern is, however, likely to be distorted by local peaks over major cities and by ridges of high land values following major roads and railways. He also cites the price of industrial land in Canada, which seems to be at a peak in Montreal and the large industrial cities of Ontario and then falls away westwards across the Prairies.

National figures however conceal a great deal of regional complexity. Smith (1969) in his book on *The North West* of Britain calculated a map of industrial land prices for the north-west, which is reproduced as Figure 5 overleaf.

This map shows that land prices in 1968 were at a peak around the cities of Manchester and Liverpool where industrial land is in very short supply, and

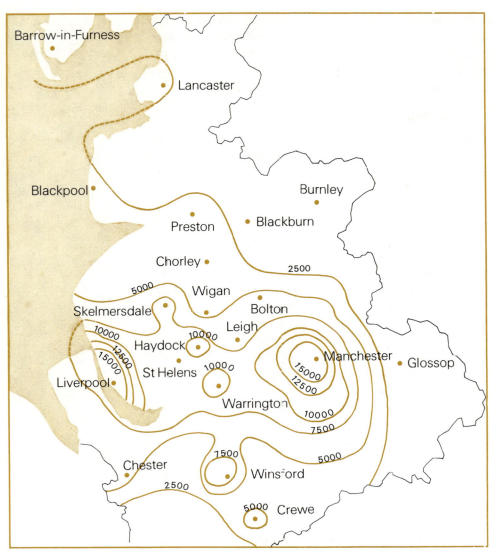

Figure 5
A contour map of the cost
of industrial land in the
north-west of England,
based on the cost of repre-
sentative industrial sites.

Source: Smith, D. M. (1969) *Industrial Britain: The North West,* Newton Abbott, David &
Charles.

that they fell away rapidly eastwards as the Pennines were approached. The
lowest prices were in Furness and in the old mill towns, while the two subsidiary
peaks at Haydock and Warrington can be explained by access to the junction
of the M6 and the East Lancs. Road for the former, and a high rate of industrial
expansion in the latter case.

## 8.2   Local variations in land costs

Maps of regional variations in land costs are as yet unfortunately rare, but
they do suggest something of the land value pattern we would expect to find
at the local level. The peaks of values around Manchester and Liverpool focus
attention on the impact of towns on the prices of industrial land. Winkler in
1957 suggested that land values would fall away from the city centre. This
stimulated a great deal of empirical investigation of land values in cities which
showed that land values in cities the size of Chicago or Rotterdam fall rapidly
away from the city centre. After about two or three miles land values begin to
level out and thereafter fall only slowly. This suggests that industrialists can
make considerable savings by avoiding sites in the city centres and locating
in suburbia.

This immediately raises the question of why, if land prices are so expensive, so many industrial concerns are found close to town centres. There are several explanations for this. Firstly, the cost of land represents a large initial cost for a newly opening factory but it represents a once and for all cost, which becomes relatively unimportant in the long term when compared with recurrent labour, raw material and transport costs. Secondly, one has to remember that land prices change through time. In particular, as cities grow, the cone of high values which is centred over the Central Business District will both increase in height and in diameter. As a result a firm now occupying a site in a high land cost area may have bought the land when it lay outside the cone of high values. Thirdly, it must be emphasized that land prices differ quite drastically within urban areas. Yeates (1965) in a study of Chicago found that local shopping centres produced localized highs in land values while major transport routes were followed by ridges of high values out from the city centre. Away from main roads and shopping centres land prices were determined by the social status of the area. In Chicago those districts bordering the lake shore had high values while the main negro areas had low values. Often values can change drastically over the distance of several streets thus making it possible for an industrial firm to occupy a low cost site only yards from the town centre.

Other local variations in land costs are likely to be due to detailed site considerations. Most manufacturing firms require level sites for their activities and many require a stable sub-soil because of the weight of the industrial equipment they need to install. In addition a very small site may attract few buyers since storage space will be scarce and working conditions cramped. Many industrial firms now seem to be moving out of the centre of British cities not because they cannot afford the costs at their existing site but because their premises are old and there is no room to rebuild on an adequate scale. A recent unpublished survey of the reasons for industrial movement in Birmingham showed that many firms relocated in order to obtain more space. While the old vacated premises remain they can often be bought or rented very cheaply by small firms, but as soon as the sites are cleared the land price alone is often much more than several decades of rent. This produces another complication to the pattern since it reveals that firms can occupy sites which they could never afford if they had to buy them at current land prices.

## 9   Conclusion

It is clear that many of the costs which an industrialist has to bear in running a factory do differ significantly from place to place. Collectively the costs discussed can be referred to as *external costs* because they all represent costs which a manufacturer cannot change. It is true that he can move his firm to a location where total costs are lower, or he can make economies by reducing the size of his labour force and increasing the amount he spends on capital equipment. What he cannot do is to change the level of wages in the area in which he happens to be located, or reduce the price of land or factory buildings which happens to prevail locally. He simply has to make the best of costs as he finds them or move to a new location. Even if he does move the success of his action will be a function of the efficiency of his decision making process, which is discussed in Unit 7 of the block. In addition whether he relocates or not his economic survival will depend not only on the level of his production costs but on geographical variations in the demand for his goods. Moreover his cost patterns and decision making are likely to be constrained by the actions of governments. Government controls on location and the geographical variations in demand form the subject of the next part of the block.

# References

ALEXANDER, J. W., BROWN, S. E. and DAHLBERG, R. E. (1958) 'Freight rates: selected aspects of uniform and nodal regions' in Karaska, G. T. and Bramhall, D. F. (eds.) (1969) *Locational Analysis for manufacturing. A selection of readings*, Cambridge, Massachusetts, M.I.T. Press.

BARLOON, M. J. (1965) 'The Interrelationship of the Changing Structure of American Transportation and Changes in Industrial Location' in Karaska, G. J. and Bramhall, D. F. (eds.) (1969) *Locational Analysis for Manufacturing. A selection of readings*, Cambridge, Massachusetts, M.I.T. Press.

BECKMANN, M. (1968) *Location Theory*, New York, Random House.

CHINITZ, B. (1960) 'The Effect of Transportation Forms on Regional Economic Growth' in Karaska, G. J. and Bramhall, D. F. (eds.) (1969) *Locational Analysis for Manufacturing. A selection of readings*, Cambridge, Massachusetts, M.I.T. Press.

CHINITZ, B. (1961) 'Contrasts in Agglomeration: New York and Pittsburgh' in Karaska, G. J. and Bramhall, D. F. (eds.) (1969) *Locational Analysis for Manufacturing. A selection of readings*, Cambridge, Massachusetts, M.I.T. Press.

FUCHS, V. R. (1967) 'Hourly Earnings Differentials by Region and Size of City' in Karaska, G. J. and Bramhall, D. F. (eds.) (1969) *Locational Analysis for Manufacturing. A selection of readings*, Cambridge, Massachusetts, M.I.T. Press.

GALBRAITH, J. K. (1967) *The New Industrial State*, Boston, Houghton Miffin.

HART, P. E. and MACBEAN, A. I. (1961) 'Regional differences in productivity, profitability and growth: a pilot study' in *Scottish Journal of Political Economy*, Vol 8, pp. 1–11.

HOLT, R. A. (1964) *The Changing Industrial Geography of the Cotton Areas of Lancashire: A study of Mill Conversion and Employment Structure*. Unpublished MA thesis, University of Manchester.

ISARD, W. and WHITNEY, V. (1950) 'Atomic power and the location of industry' in Karaska, G. T. and Bramhall, D. F. (eds.) (1969) *Locational Analysis for Manufacturing. A selection of readings*, Cambridge, Massachusetts, M.I.T. Press.

KRUTILLA, J. V. (1965) 'Locational factors influencing recent aluminium expansion' in *Southern Economic Journal*, Vol. 21, pp. 273–88.

LESER, C. E. V. (1949) 'Changes in level and diversity of employment in the regions of Great Britain 1939–47' in *Economic Journal*, Vol. 59.

LESER, C. E. V. (1950) 'Output per head in different parts of the United Kingdom' in *Journal of the Royal Statistical Society*, Series A, Vol 113, pp. 207–19.

OLIVER, F. R. (1964) 'Inter-regional migration and unemployment 1951–61' in *Journal of the Royal Statistical Society*, Series A, Vol. 127, pp. 42–69.

SINGER, H. W. and LESER, C. E. V. (1948) 'Industrial productivity in England and Scotland' in *Journal of the Royal Statistical Society*, Series A, Vol. 111, pp. 309–30.

SMITH, D. M. (1969) *Industrial Britain: The North West*, Newton Abbott, David & Charles.

SMITH, D. M. (1971) *Industrial Location: An economic geographical analysis*, New York and Toronto, John Wiley & Sons.

THIRLWALL, A. P. (1966) 'Regional unemployment as a cyclical phenomena' in *Scottish Journal of Political Economy*, Vol. 13, June 1966, pp. 205–19.

US DEPARTMENT OF LABOUR (1952) 'The mobility of Tool and Die Makers, 1940–51' in *Bureau of Labour Statistics Bulletin*, No. 1120, 14 November 1952.

VYVER, F. T. DE (1951) 'Labour factors in the industrial development of the South' in Karaska, G. T. and Bramhall, D. F. (eds.) (1969) *Locational Analysis for Manufacturing. A selection of readings*, Cambridge, Massachusetts, M.I.T. Press.

YEATES, M. H. (1965) 'Some factors affecting the spatial distribution of Chicago land values, 1910–1960 in *Economic Geography*, Vol. 41, pp. 57–70.

**Notes**

**Notes**

**Notes**

# New Trends in Geography